ENERGY, ELECTRICITY AND NUCLEAR POWER ESTIMATES FOR THE PERIOD UP TO 2050

REFERENCE DATA SERIES No. 1

ENERGY, ELECTRICITY AND NUCLEAR POWER ESTIMATES FOR THE PERIOD UP TO 2050

2022 Edition

INTERNATIONAL ATOMIC ENERGY AGENCY
VIENNA, 2022

ENERGY, ELECTRICITY AND
NUCLEAR POWER ESTIMATES
FOR THE PERIOD UP TO 2050
IAEA-RDS-1/42
ISBN 978-92-0-136722-8
ISSN 1011-2642
Printed by the IAEA in Austria
September 2022
Cover photo credit:
Tapani Karjanlahti / TVO 2022

CONTENTS

Introduction

Reference Data Series No. 1 (RDS-1) is an annual publication — currently in its 42nd edition — containing estimates of energy, electricity and nuclear power trends up to the year 2050.

The publication is organized into world and regional subsections and starts with a summary of the status of nuclear power in IAEA Member States as of the end of 2021 based on the latest statistical data collected by the IAEA's Power Reactor Information System. It then presents global and regional projections for energy and electricity up to 2050 derived from two international studies: the International Energy Agency's World Energy Outlook 2021 [1] and the United States Energy Information Administration's International Energy Outlook 2021 [2]. The energy and electricity data for 2021 are estimated, as the latest information available from the United Nations Department of Economic and Social Affairs [3] and International Energy Agency [4] is for 2019. Population data originate from World Population Prospects 2022 [5], published by the Population Division of the United Nations Department of Economic and Social Affairs.

Global and regional nuclear power projections are presented as low and high cases, encompassing the uncertainties inherent in projecting trends. The projections are based on a critical review of (i) the global and regional energy, electricity and nuclear power projections made by other international organizations, (ii) national projections supplied by individual countries for a recent joint OECD Nuclear Energy Agency and IAEA study [6] and (iii) estimates of the expert group participating in an annual IAEA consultancy meeting.

The nuclear electrical generating capacity estimates presented in Table 5 on page 24 of the publication are derived using a country by country 'bottom-up' approach. In deriving these estimates, the group of experts considered all operating reactors, possible licence renewals, planned shutdowns and plausible construction projects foreseen for the next several decades. The experts build the estimates project by project by assessing the plausibility of each considering a high and low case.

The assumptions of the low case are that current market, technology and resource trends continue and there are few additional changes in explicit laws, policies and regulations affecting nuclear power. This case was designed to produce a 'conservative but plausible' set of projections. Additionally, the low case does not assume that targets for nuclear power in a particular country will necessarily be achieved. The high case projections are much more ambitious but are still plausible and technically feasible. Country policies on climate change are also considered in the high case. In both cases the same outlook of economic and electricity demand growth based on current expectations is assumed. The high case projection is not intended to reflect a net zero carbon emissions ambition. It does not assume a specific pathway for energy system transitions in the different countries but integrates the expressed intentions of the countries for expanding the use of nuclear power.

The low and high estimates reflect contrasting, but not extreme, underlying assumptions about the different driving factors that have an impact on nuclear power deployment. These factors, and the way they might evolve, vary from country to country. The estimates presented provide a plausible range of nuclear capacity development by region and worldwide. They are not intended to be predictive nor to reflect the whole range of possible futures from the lowest to the highest feasible.

By 2050 global final energy consumption is projected to increase by about 30% and electricity production is expected to double [1,2]. Worldwide, coal remains the dominant energy source for electricity production at about 36% for 2021. While its share in electricity production has changed little since 1980, that of nuclear, renewables and natural gas has increased over the past 40 years. Today, nuclear contributes about 10% of global electricity production.

The adoption of the Glasgow Climate Pact following the 26th United Nations Climate Change Conference of the Parties (COP26) in November 2021 has led to renewed momentum toward reaching net zero global CO_2 emissions by 2050. In the lead-up to COP26, a number of countries revised their nationally determined contributions, committed to reaching net zero CO_2 emissions in the coming decades and recognized the role that nuclear energy can play in reaching this climate goal. One of the key outcomes of COP26 is the pledge by a number of countries and international finance institutions to stop financing new coal power plants and to phase out existing coal power plants.

Energy security and resilience are currently major policy concerns. Recent events such as the COVID-19 pandemic, geopolitical tensions and military conflict in Europe have impacted the reliability of energy systems, impeded energy flows across regions and led to significant increases in energy prices. There is growing recognition of the role of nuclear energy as a key contributor to the security of energy supply to avert future energy supply and price shocks.

In light of this evolving energy landscape, with strong commitment to climate action and renewed scrutiny of energy supply security, a number of Member States have revised their national energy policy, leading to decisions for the long term operation of existing reactors, new construction of Generation III/III+ designs, and the development and deployment of small modular reactors.

These factors are contributing to government announcements of a larger role for nuclear energy in their energy and climate strategies, leading to a notable upward revision of the high case by about 10% compared with the 2021 edition of this publication. Relative to a global nuclear electrical generating capacity of 390 gigawatts (electrical) (GW(e)) in 2021, the low case projections indicate that world nuclear capacity will remain essentially the same at 404 GW(e). In the high case, world nuclear capacity is expected to more than double to 873 GW(e)[1] by 2050.

There are a number of necessary conditions for a substantial increase in installed nuclear capacity. A number of these issues are being addressed, including international efforts toward regulatory and industrial harmonization, as well as progress with final disposal of high level radioactive waste. However, a number of challenges remain, including financing, economic and supply chain difficulties for new nuclear construction in some regions.

Climate change mitigation is a key driver of decisions to continue or expand the use of nuclear power. According to the IAEA [7], the use of nuclear power has avoided about 70 gigatonnes of CO_2 emissions over the past 50 years. Commitments made under the Paris Agreement and other initiatives could support nuclear power development, provided the necessary energy policies and market designs are established to facilitate investments in dispatchable low carbon technologies.

As stated by the International Energy Agency [8], almost half of the CO_2 emission reductions needed to reach net zero in 2050 will need to come from technologies that are currently under development but are not yet on the market. This is true for nuclear technologies such as small and medium sized, modular and

---

[1] Owing to the uncertain situation in some specific countries in Eastern Europe, the expert group decided not to revise projections for those countries.

other advanced reactors. Accelerating the pace of innovation and demonstration of these technologies is required if nuclear is to play a role in decarbonization beyond electricity by providing low carbon heat or hydrogen to the industrial and transport sectors.[2]

Currently, about two thirds of nuclear power reactors have been in operation for over 30 years, highlighting the need for significant new nuclear capacity to offset retirements in the long term. Uncertainty remains regarding the replacement of the large number of reactors scheduled to be retired by about 2030 and beyond, particularly in Northern America. However, ageing management programmes and long term operation are being implemented for an increasing number of reactors. Additionally, new policy measures are being implemented to support the competitiveness of existing reactors in liberalized electricity markets.

It is important to consider the changes in nuclear electrical generating capacity in each region within the context of region specific factors. In recent years, construction cost overruns and delays for first of a kind projects have led to high project risk perception in the Americas and Europe, hampering investment decisions for new projects. In some regions nuclear power plants have been built on time and on budget. The expert group assumed that the aforementioned challenges may continue to affect some nuclear development plans.

The current pace of nuclear power development shows that urgent actions are needed to maintain the existing role of nuclear power in the energy mix. The involvement of a broad range of actors including policy makers, the nuclear industry and international organizations, along with active engagement with the public, is necessary.

The underlying fundamentals of population and electricity consumption growth, as well as concerns about climate change and air quality concerns, the security of energy supply and the price volatility of other fuels, point to nuclear energy continuously playing an essential role in the energy mix in the longer run, provided concerted actions are taken.

––––––––––––

[2] The projections do not explicitly take into account all the potential technologies (small and advanced reactors) and potential uses of nuclear power (e.g. heat, hydrogen, water desalination) under climate change constraints.

Geographical Regions

The nuclear electrical generating capacity projections presented in RDS-1 are grouped according to the geographical regions used by the Statistics Division of the United Nations Secretariat (see annex I to Ref. [9]). The designations employed and the presentation of material in this publication do not imply the expression of any opinion whatsoever on the part of the IAEA concerning the legal status of any country, territory, city or area or of its authorities, or concerning the delimitation of its frontiers or boundaries.

Notes

The estimates for nuclear electricity production in 2021 are from the 2022 edition of Nuclear Power Reactors in the World, Reference Data Series No. 2 (RDS-2) [10]. The estimates for energy and electricity are made by the IAEA Secretariat on the basis of different international and national data sources available as of July 2022.

In accordance with the International Recommendations for Energy Statistics [11], the estimates for the breakdown of historical electricity production by energy source are expressed in gross figures. Gross electricity production is the total electrical energy produced by all generating units and installations measured at the output terminals of the generators. Current data on nuclear electrical production and future estimates of nuclear and total electrical production are expressed in net values, as the data are adapted from the RDS-2 publication.

Owing to rounding, numbers presented throughout this publication may not add up precisely to the totals provided, and percentages may not precisely reflect the absolute figures.

Total final energy consumption refers to all fuel and energy delivered to end users for their energy use.

Nuclear electrical generating capacity estimates consider the scheduled retirement of older units at the end of their lifetime.

The global and regional nuclear electrical production data and the nuclear electrical generating capacity data cannot be used to calculate average annual capacity factors for nuclear power plants, as the nuclear electrical generating capacity data are year-end capacity.

World

7 877

million people

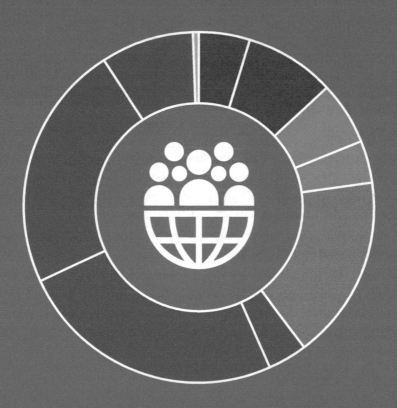

Energy Overview 2021

 19.5%
of final energy consumed was electricity

 27 007 TW·h
of electricity produced

 **9.8%**
of electricity produced by nuclear

Nuclear Power Development in 2021

- At the end of 2021, 437 nuclear power reactors were operational, with a total net installed power capacity of 389.5 GW(e).

- In addition, 56 reactors with a total capacity of 58.1 GW(e) were under construction.

- Six new nuclear power reactors with a total capacity of 5.2 GW(e) were connected to the grid, and ten reactors with a total capacity of 8.7 GW(e) were retired. Construction began on ten new reactors that are expected to add a total capacity of 8.8 GW(e).

- Compared with 2020, total electricity production from all energy sources increased by 7% and electricity production from nuclear power reactors increased about 4% to 2 653 TW·h.

- Nuclear power accounted for 9.8% of total electricity production in 2021, a decrease of 0.4 percentage points from the previous year.

- The reduction in global electricity demand in 2020 was the biggest annual decline since the mid-20th century. In 2021 global electricity consumption rebounded and exceeded levels for 2019. The total energy consumption increased but did not reach the level of 2019.

TABLE 1. NUCLEAR POWER REACTORS IN THE WORLD (end of 2021)

Country	Operational		Under Construction		Nuclear Electricity Production in 2021	
	Number of units	Net capacity (MW(e))	Number of units	Net capacity (MW(e))	TW·h	% of total
World Total[a]	437	389 508	56	58 096	2653.1	9.8
Argentina	3	1 641	1	25	10.2	7.2
Armenia	1	448			1.9	25.3
Bangladesh			2	2 160		
Belarus	1	1 110	1	1 110	5.4	14.1
Belgium	7	5 942			48.0	50.8
Brazil	2	1 884	1	1 340	13.9	2.4
Bulgaria	2	2 006			15.8	34.6
Canada	19	13 624			86.8	14.3
China	53	50 034	16	15 967	383.2	5.0
Czech Republic	6	3 934			29.0	36.6
Finland	4	2 794	1	1 600	22.6	32.8
France	56	61 370	1	1 630	363.4	69.0
Germany	3	4 055			65.4	11.9
Hungary	4	1 916			15.1	44.7
India	22	6 795	8	6 028	39.8	2.8

Country						
Iran, Islamic Republic of	1	915	1	974	3.2	1.0
Japan	33	31 679	2	2 653	61.3	5.1
Korea, Republic of	24	23 091	4	5 360	150.5	26.5
Mexico	2	1 552			11.6	3.4
Netherlands	1	482			3.6	3.1
Pakistan	5	2 242	1	1 014	15.8	10.6
Romania	2	1 300			10.4	19.0
Russian Federation	37	27 727	4	3 759	208.4	19.2
Slovakia	4	1 868	2	880	14.6	52.3
Slovenia	1	688			5.4	36.9
South Africa	2	1 854			12.2	3.4
Spain	7	7 121			54.2	20.8
Sweden	6	6 882			51.4	31.4
Switzerland	4	2 960			18.6	29.5
Türkiye			3	3 342		
Ukraine	15	13 107	2	2 070	81.1	55.0
United Arab Emirates	2	2 762	2	2 690	10.1	1.3
United Kingdom	12	7 343	2	3 260	41.8	14.8
United States of America	93	95 523	2	2 234	771.6	19.6

[a] Includes the following data from Taiwan, China: 3 units in operation with a total capacity of 2859 MW(e) and 26.8 TW·h of nuclear electricity generation, representing 9.5% of the total electricity produced.

FIGURE 1. WORLD NUCLEAR ELECTRICITY PRODUCTION IN 2021

Note: The nuclear electricity production in Taiwan, China, was 26.8 TW·h.

FIGURE 2. SHARE OF NUCLEAR IN TOTAL ELECTRICITY PRODUCTION IN THE WORLD IN 2021

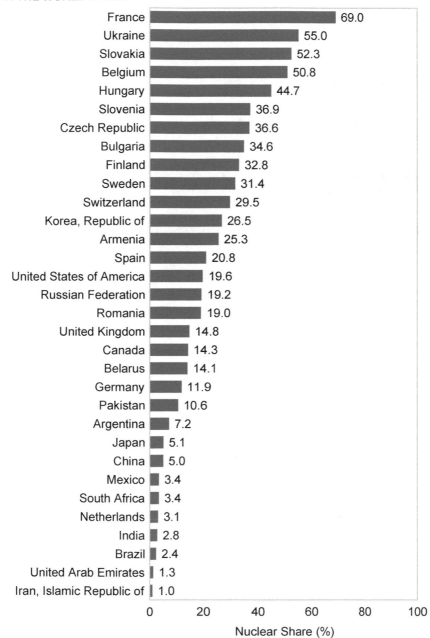

Country	Nuclear Share (%)
France	69.0
Ukraine	55.0
Slovakia	52.3
Belgium	50.8
Hungary	44.7
Slovenia	36.9
Czech Republic	36.6
Bulgaria	34.6
Finland	32.8
Sweden	31.4
Switzerland	29.5
Korea, Republic of	26.5
Armenia	25.3
Spain	20.8
United States of America	19.6
Russian Federation	19.2
Romania	19.0
United Kingdom	14.8
Canada	14.3
Belarus	14.1
Germany	11.9
Pakistan	10.6
Argentina	7.2
Japan	5.1
China	5.0
Mexico	3.4
South Africa	3.4
Netherlands	3.1
India	2.8
Brazil	2.4
United Arab Emirates	1.3
Iran, Islamic Republic of	1.0

Nuclear Share (%)

Note: The share of nuclear in the total electricity production of Taiwan, China, was 9.5%.

13

FIGURE 3. WORLD FINAL ENERGY CONSUMPTION BY ENERGY SOURCE

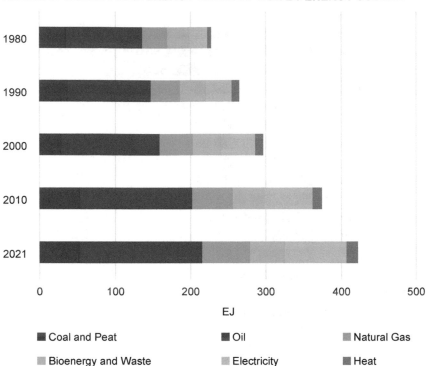

Final Energy Consumption

Since 1980 fossil fuels have continued to dominate final energy consumption, although there has been a gradual reduction in their combined share from 74% in 1980 to 66% in 2021.

The share of coal declined slightly from 1980 to 2000 and increased from 2000 to 2010 and has since declined again. Natural gas has maintained a consistent share of about 15%. The share of oil has declined slightly since 1980, stabilizing at about 40% since 2010.

The share of electricity has undergone the most significant change since 1980, increasing by 9 percentage points, with consumption growing at an average annual rate of about 3%.

Looking to the future, electricity consumption is expected to increase faster than final energy consumption, thus it is anticipated that the share of electricity will continue to grow.

FIGURE 4. WORLD TOTAL ELECTRICITY PRODUCTION
BY ENERGY SOURCE

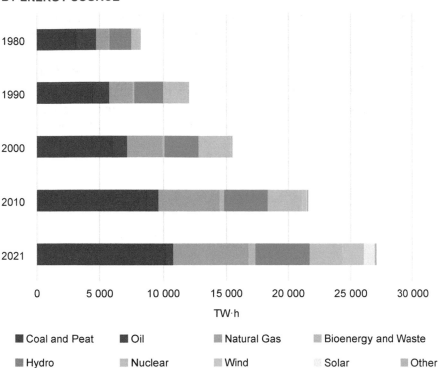

Electricity Production

With a share of more than 60%, fossil fuels — particularly coal — have remained dominant sources of electricity production since 1980, despite increases in the combined share of nuclear and renewables over the years.

The share of natural gas has increased more than 10 percentage points since 1980. The share of coal remained around 40% until 2010 but has since gradually decreased by a few percentage points. Of all fossil fuels, the share of oil has experienced the most significant change, decreasing from about 20% in 1980 to about 2% in 2021.

Hydro remains the largest contributor of low carbon electricity, accounting for 16%, although its share has decreased by about 4 percentage points since 1980. In recent years, the share of solar and wind has undergone a rapid increase, rising from less than 1% in 1980 to 9% in 2021.

The share of nuclear grew rapidly from 1980 to 1990, almost doubling, but has declined since 2000.

Energy and Electricity Projections

- Final energy consumption is expected to increase by about 12% from 2021 levels by 2030 and by about 27% by 2050, at an average annual rate of approximately 1%.

- Electricity consumption is expected to grow at a faster rate of about 2.4% per year. Electricity consumption is expected to double by 2050.

- By 2050 the share of electricity in final energy consumption is expected to increase by about 10 percentage points from its 2021 share.

FIGURE 5. WORLD FINAL CONSUMPTION OF ENERGY AND ELECTRICITY

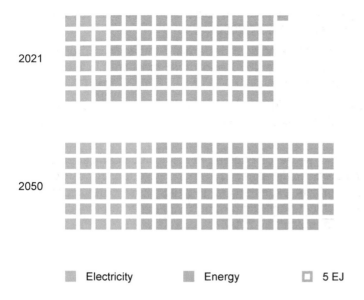

| | Electricity | | Energy | | 5 EJ |

TABLE 2. WORLD FINAL CONSUMPTION OF ENERGY AND ELECTRICITY, EJ

Final Consumption	2021	2030	2040	2050
Energy	422.4	471.2	498.3	535.1
Electricity	82.3	105.5	132.1	159.6
Electricity as % of Energy	*19.5%*	*22.4%*	*26.5%*	*29.8%*

Nuclear Electrical Generating Capacity Projections

- Total electrical generating capacity is expected to increase by about 23% by 2030 and then double by 2050.

- In the high case, nuclear electrical generating capacity is projected to increase by about 23% by 2030 and more than double by 2050 compared with 2021 capacity.

- In the low case, nuclear electrical generating capacity is projected to decline by about 2% by 2030 and then increase by about 3.5% by 2050.

- In the low case, the share of nuclear in total electrical generating capacity is projected to decrease by 2050. A reduction of about 2.4 percentage points is expected. In the high case, the share of nuclear in total electrical generating capacity is expected to increase by half of a percentage point by 2050.

FIGURE 6. WORLD NUCLEAR ELECTRICAL GENERATING CAPACITY

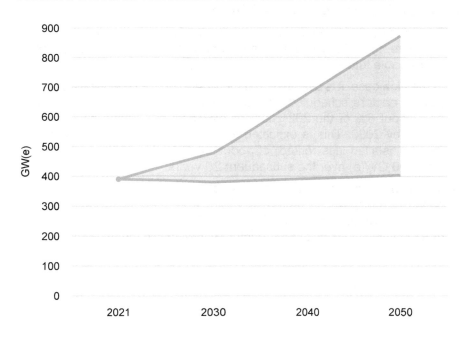

TABLE 3. WORLD TOTAL AND NUCLEAR ELECTRICAL GENERATING CAPACITY, GW(e)

Electrical Capacity	2021	2030		2040		2050	
		Low	High	Low	High	Low	High
Total	8 208	10 079	10 079	12 841	12 841	16 590	16 590
Nuclear	390	381	479	392	676	404	873
Nuclear as % of Electrical Capacity	*4.8%*	*3.8%*	*4.8%*	*3.1%*	*5.3%*	*2.4%*	*5.3%*

Reactor Retirements and Additions

- Two out of every three nuclear power reactors have been in operation for more than 30 years and are scheduled for retirement in the foreseeable future.

- In the high case, it is assumed that the operating life of several nuclear power reactors scheduled for retirement will be extended such that only about 8% of the 2021 nuclear electrical generating capacity is retired by 2030. This is expected to result in net capacity additions (newly installed less retired) of about 90 GW(e) by 2030 and more than 390 GW(e) over the subsequent 20 years.

- In the low case, it is assumed that about 18% of existing nuclear power reactors will be retired by 2030, while new reactors will add about 60 GW(e) of capacity. Between 2030 and 2050 it is expected that capacity additions of new reactors will slightly exceed retirements.

FIGURE 7. WORLD NUCLEAR CAPACITY: ACTUAL, RETIREMENTS AND ADDITIONS

HIGH CASE

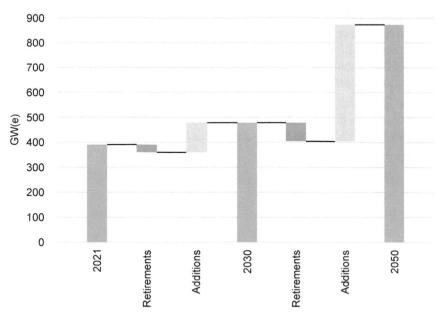

LOW CASE

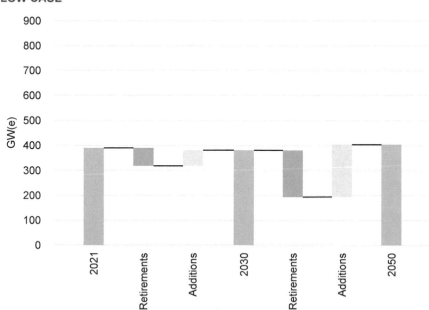

Electricity and Nuclear Production Projections

- Total electricity production is expected to increase by about 23% by 2030 and by 85% by 2050 compared with 2021 levels.

- In the high case, nuclear electricity production is expected to increase by about 40% from the 2021 level by 2030 and by more than 2.5-fold by 2050. The share of nuclear in total electricity production is expected to increase by more than 3 percentage points.

- In the low case, nuclear electricity production is expected to increase by about 12% from the 2021 level by 2030, rising to 29% by 2050. The share of nuclear in total electricity production is expected to decline by about 3 percentage points.

FIGURE 8. WORLD NUCLEAR ELECTRICITY PRODUCTION

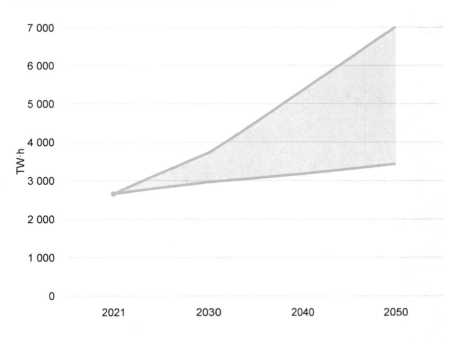

TABLE 4. WORLD TOTAL AND NUCLEAR ELECTRICAL PRODUCTION, TW·h

Electricity Production	2021	2030		2040		2050	
		Low	High	Low	High	Low	High
Total	27 007	33 275	33 275	41 508	41 508	50 071	50 071
Nuclear	2 653	2 963	3 724	3 169	5 336	3 435	7 010
Nuclear as % of Electricity Production	*9.8%*	*8.9%*	*11.2%*	*7.6%*	*12.9%*	*6.9%*	*14.0%*

TABLE 5. WORLD NUCLEAR ELECTRICAL GENERATING CAPACITY, GW(e)

Region	2021	2030 Low	2030 High	2040 Low	2040 High	2050 Low	2050 High
World Total	389.5	381	479	392	676	404	873
Northern America	109.1	86	110	63	110	40	124
Latin America and the Caribbean	5.1	6	6	9	16	12	25
Northern, Western and Southern Europe	99.6	84	90	70	112	43	133
Eastern Europe	53.0	53	62	55	93	63	104
Africa	1.9	2	3	8	12	9	21
Western Asia	3.2	8	9	12	19	14	24
Southern Asia	10.0	18	29	32	49	47	78
Central and Eastern Asia	107.7	124	170	142	260	173	347
South-eastern Asia				1	5	3	15
Oceania							2

TABLE 6. WORLD NUCLEAR ELECTRICITY PRODUCTION, TW·h

Region	2021	2030		2040		2050	
		Low	High	Low	High	Low	High
World Total	2653.1	2 963	3 724	3 169	5 336	3 435	7 010
Northern America	858.4	699	896	515	908	326	1 031
Latin America and the Caribbean	35.7	42	46	65	123	92	197
Northern, Western and Southern Europe	674.4	678	727	580	928	353	1 092
Eastern Europe	379.8	399	465	425	707	492	816
Africa	12.2	13	22	57	84	69	151
Western Asia	12.0	56	64	91	141	112	189
Southern Asia	58.8	132	211	239	361	370	609
Central and Eastern Asia	621.8	943	1 293	1 188	2 047	1 597	2 793
South-eastern Asia				8	37	24	118
Oceania							14

Northern America

375
million people

Energy Overview 2021

 22.0%
of final energy consumed was electricity

 4 800TW·h
of electricity produced

 **17.9**%
of electricity produced by nuclear

FIGURE 9. FINAL ENERGY CONSUMPTION BY ENERGY SOURCE IN THE NORTHERN AMERICA REGION

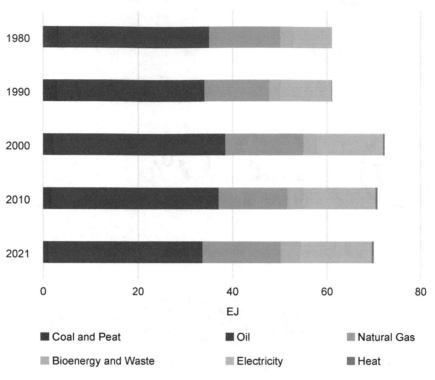

Final Energy Consumption

Since 1980, the share of fossil fuels in final energy consumption has remained above 70%, with a slight reduction from 82% in 1980 to 72% in 2021.

Of all fossil fuels, oil has the largest share, having remained at about 50% since 1980. In 2021 the share of oil decreased to 47%.

With a share of 24%, natural gas was the second largest energy source in 2021. Its share has remained relatively stable since 1980.

From 1980 to 2010, the share of electricity gradually increased by 8 percentage points. Its share in 2021 was 22% of final energy consumption.

FIGURE 10. ELECTRICITY PRODUCTION BY ENERGY SOURCE IN
THE NORTHERN AMERICA REGION

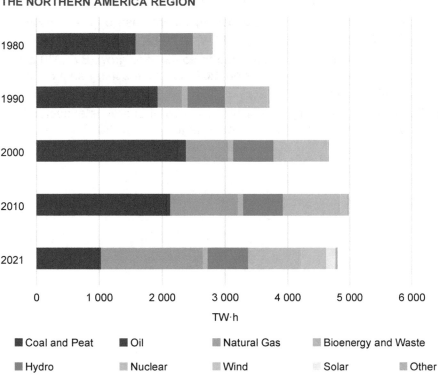

Electricity Production

Fossil fuels contributed more than half of the electricity produced in 2021.

The share of coal has decreased by more than half since 1980, whereas the share of natural gas has more than doubled. The share of oil has decreased from 10% in 1980 to around 1% in 2021.

Nuclear is the largest low carbon energy source. Its share nearly doubled from 1980 to 1990 and has remained stable at about 18% since 1990.

The share of hydro has decreased by about 6 percentage points over the past 40 years.

The share of wind has rapidly increased since 2000, exceeding 8% by 2021. In recent years, the share of solar has also undergone a rapid increase, rising from less than 1% in 2010 to 3% in 2021.

Energy and Electricity Projections

- Final energy consumption is expected to decrease by about 5% by 2040 and then stay almost constant up to 2050.

- Electricity consumption is expected to continue to grow. By 2030 it is projected to increase by 11% from 2021 levels, reaching an increase of about 44% by 2050.

- The share of electricity in final consumption of energy is expected to gradually increase by about 11 percentage points by 2050.

FIGURE 11. FINAL CONSUMPTION OF ENERGY AND ELECTRICITY
IN THE NORTHERN AMERICA REGION

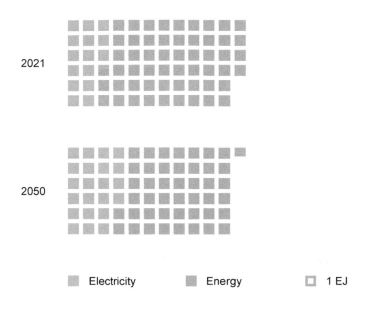

2021

2050

Electricity   Energy   1 EJ

TABLE 7. FINAL CONSUMPTION OF ENERGY AND ELECTRICITY
IN THE NORTHERN AMERICA REGION, EJ

Final Consumption	2021	2030	2040	2050
Energy	70.0	68.3	66.9	66.8
Electricity	15.4	17.1	19.5	22.1
Electricity as % of Energy	*22.0%*	*25.0%*	*29.1%*	*33.1%*

Nuclear Electrical Generating Capacity Projections

- Total electrical generating capacity is projected to increase by almost 5% by 2030 and by about 40% by 2050.

- A significant reduction in nuclear electrical generating capacity is projected over the next three decades for the low case, whereas the high case is expected to remain relatively stable until 2040 with a considerable increase by 2050.

- In the high case, nuclear electrical generating capacity is projected to remain roughly constant until 2040, with an increase of about 14% by 2050. The share of nuclear in total electrical capacity is expected to remain stable until 2030 and then decrease by more than 1 percentage point by 2050.

- In the low case, nuclear electrical generating capacity is projected to decrease by about 20% from current levels by 2030 and to be around one third of current capacity by 2050. The share of nuclear in total electrical capacity is projected to decrease by about 2 percentage points by 2030 and by almost 6 percentage points by 2050.

FIGURE 12. NUCLEAR ELECTRICAL GENERATING CAPACITY
IN THE NORTHERN AMERICA REGION

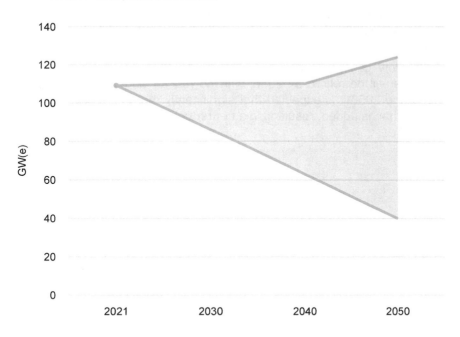

TABLE 8. TOTAL AND NUCLEAR ELECTRICAL GENERATING CAPACITY
IN THE NORTHERN AMERICA REGION, GW(e)

Electrical Capacity	2021	2030		2040		2050	
		Low	High	Low	High	Low	High
Total	1 428	1 495	1 495	1 628	1 628	1 990	1 990
Nuclear	109	86	110	63	110	40	124
Nuclear as % of Electrical Capacity	*7.6%*	*5.8%*	*7.4%*	*3.9%*	*6.8%*	*2.0%*	*6.2%*

Reactor Retirements and Additions

- In the high case, a net increase in nuclear capacity of 1 GW(e) is expected by 2030. Between 2030 and 2050 it is expected that capacity additions of new reactors will exceed retirements by 14 GW(e).

- In the low case, it is assumed that about 20% of nuclear power reactors will be retired by 2030, with no reactor additions. Between 2030 and 2050 it is expected that significantly more capacity will be retired than is added, resulting in a net reduction in capacity of almost 50 GW(e).

FIGURE 13. NUCLEAR CAPACITY IN THE NORTHERN AMERICA REGION:
ACTUAL, RETIREMENTS AND ADDITIONS

HIGH CASE

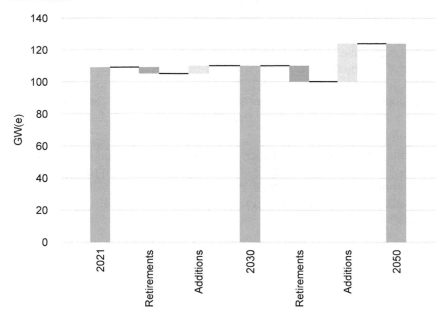

LOW CASE

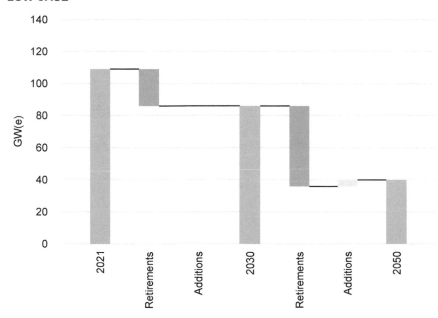

Electricity and Nuclear Production Projections

- Total electricity production is projected to increase about 11% by 2030 and is expected to be about 43% higher than 2021 production levels by 2050.

- In the high case, nuclear electricity production is projected to increase about 4.5% by 2030 and is expected to be about 20% higher than 2021 production levels by 2050. The share of nuclear in total electricity production is expected to decrease, but less than in the low case, with a reduction of about 1 percentage point by 2030 and almost 3 percentage points by 2050.

- In the low case, nuclear electricity production is projected to decrease by 19% from 2021 levels by 2030 and by almost two thirds by 2050. The share of nuclear in total electricity production is expected to decrease by about 5 percentage points by 2030 and 13 percentage points by 2050.

**FIGURE 14. NUCLEAR ELECTRICITY PRODUCTION
IN THE NORTHERN AMERICA REGION**

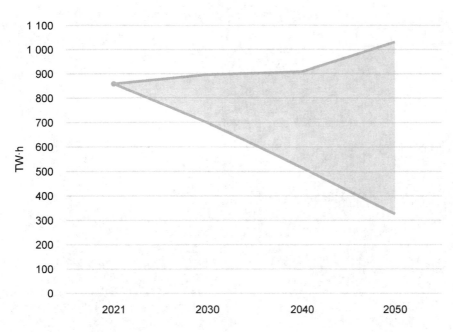

**TABLE 9. TOTAL AND NUCLEAR ELECTRICAL PRODUCTION
IN THE NORTHERN AMERICA REGION, TW·h**

Electricity Production	2021	2030		2040		2050	
		Low	High	Low	High	Low	High
Total	4 800	5 334	5 334	6 091	6 091	6 883	6 883
Nuclear	858	699	896	515	908	326	1 031
Nuclear as % of Electricity Production	17.9%	13.1%	16.8%	8.5%	14.9%	4.7%	15.0%

Latin America and the Caribbean

654
million people

Energy Overview 2021

 20.0%
of final energy consumed was electricity

 1 679TW·h
of electricity produced

 **2.1**%
of electricity produced by nuclear

FIGURE 15. FINAL ENERGY CONSUMPTION BY ENERGY SOURCE
IN THE LATIN AMERICA AND THE CARIBBEAN REGION

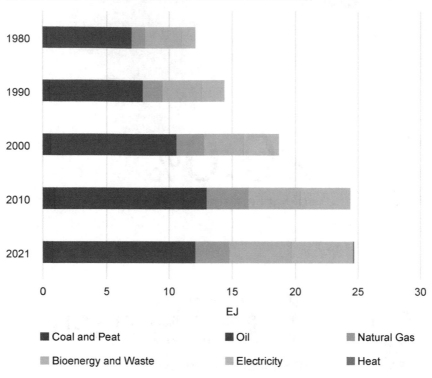

Final Energy Consumption

From 1980 to 2000 the share of fossil fuels in final energy consumption increased slightly, but since 2000 it has gradually decreased. The combined share of fossil fuels in 2021 was about 60%.

Oil accounted for about 46% of final energy consumption in 2021, although its share has decreased by about 9 percentage points since 1980.

The share of natural gas has risen by a few percentage points over the past 40 years but declined in 2021.

The share of coal has remained small at about 3%.

The share of electricity has undergone the most significant change, more than doubling since 1980.

FIGURE 16. ELECTRICITY PRODUCTION BY ENERGY SOURCE IN
THE LATIN AMERICA AND THE CARIBBEAN REGION

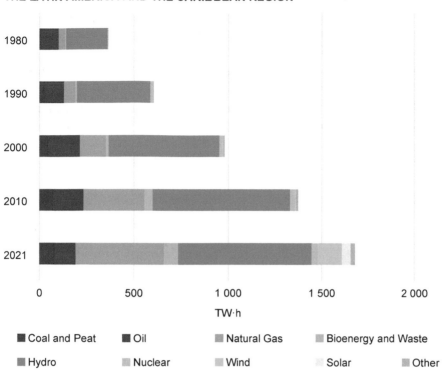

Electricity Production

Since 1980 hydro has been the largest source of electricity; its production has increased throughout this period, although its share has decreased by about 18 percentage points. In 2000, other renewables started contributing to electricity production. In 2021 the combined share of other renewables exceeded 16%.

Of all fossil fuels, natural gas accounted for the largest share of electricity production in 2021, having displaced oil as the largest source after 2000. The share of natural gas has undergone an almost threefold increase over the past 40 years. The share of coal has more than doubled since 1980, whereas the share of oil has steadily decreased by almost 20 percentage points.

The share of nuclear has increased almost fourfold since 1980, although its overall share has remained relatively small and was just over 2% in 2021.

Energy and Electricity Projections

- Final consumption of energy is expected to increase by about 21% from 2021 levels by 2030 and by 40% by 2050, at an average annual rate of about 1.2%.

- Electricity consumption is expected to grow at a faster rate of about 3% per year, more than doubling over the next 29 years.

- By 2050 the share of electricity in final energy consumption is expected to increase by about 11 percentage points from its 2021 share.

FIGURE 17. FINAL CONSUMPTION OF ENERGY AND ELECTRICITY
IN THE LATIN AMERICA AND THE CARIBBEAN REGION

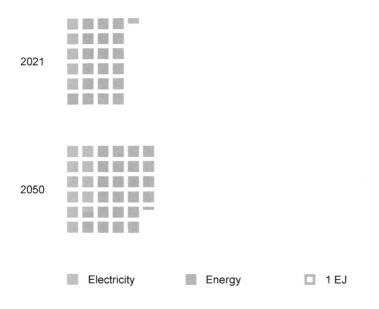

TABLE 10. FINAL CONSUMPTION OF ENERGY AND ELECTRICITY
IN THE LATIN AMERICA AND THE CARIBBEAN REGION, EJ

Final Consumption	2021	2030	2040	2050
Energy	24.5	29.7	31.9	34.3
Electricity	4.9	6.3	8.3	10.5
Electricity as % of Energy	*20.0%*	*21.2%*	*26.0%*	*30.6%*

Nuclear Electrical Generating Capacity Projections

- Total electrical generating capacity is projected to increase by about 19% by 2030 and to almost double by 2050.

- In the high case, nuclear electrical generating capacity is projected to increase fivefold by 2050, with its share of total electrical capacity growing by 1.6 percentage points.

- In the low case, nuclear electrical generating capacity is projected to triple over the next 29 years, although its share in total electrical capacity is expected to remain nearly constant.

FIGURE 18. NUCLEAR ELECTRICAL GENERATING CAPACITY
IN THE LATIN AMERICA AND THE CARIBBEAN REGION

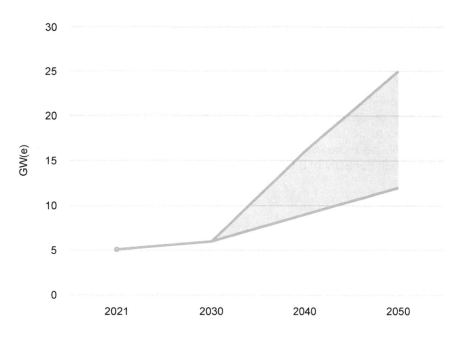

TABLE 11. TOTAL AND NUCLEAR ELECTRICAL GENERATING CAPACITY
IN THE LATIN AMERICA AND THE CARIBBEAN REGION, GW(e)

Electrical Capacity	2021	2030		2040		2050	
		Low	High	Low	High	Low	High
Total	495	589	589	731	731	959	959
Nuclear	5.1	6	6	9	16	12	25
Nuclear as % of Electrical Capacity	*1.0%*	*1.0%*	*1.0%*	*1.2%*	*2.2%*	*1.3%*	*2.6%*

Reactor Retirements and Additions

- In the high case, no reactor retirements are expected by 2030 and it is assumed that just over 1 GW(e) of capacity will be added. Between 2030 and 2050 it is expected that there will be a significant number of additions with only a few retirements, resulting in a net increase in capacity of almost 19 GW(e).

- In the low case, it is assumed that there will be a net increase in capacity of about 1 GW(e) by 2030 as well, and no reactors are expected to be retired. Between 2030 and 2050 it is expected that there will be more capacity added than retired, resulting in a net increase in capacity of 6 GW(e).

FIGURE 19. NUCLEAR CAPACITY IN THE LATIN AMERICA AND THE
CARIBBEAN REGION: ACTUAL, RETIREMENTS AND ADDITIONS

HIGH CASE

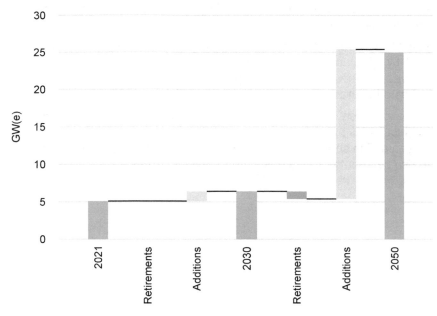

LOW CASE

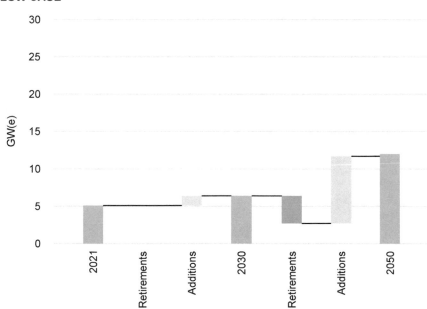

Electricity and Nuclear Production Projections

- Total electricity production is projected to rise significantly by 2030, with an increase of about 25% from 2021 levels. It is expected to more than double by 2050.

- In the high case, nuclear electricity production is projected to rise about 28% by 2030 and to increase more than 5.5-fold over the subsequent 20 years. The share of nuclear in total electricity production is expected to gradually increase, nearly tripling by 2050.

- In the low case, nuclear electricity production is projected to increase by about 17% by 2030 and to more than double over the subsequent 20 years. The share of nuclear in total electricity production is expected to rise by about half of a percentage point by 2050.

FIGURE 20. NUCLEAR ELECTRICITY PRODUCTION
IN THE LATIN AMERICA AND THE CARIBBEAN REGION

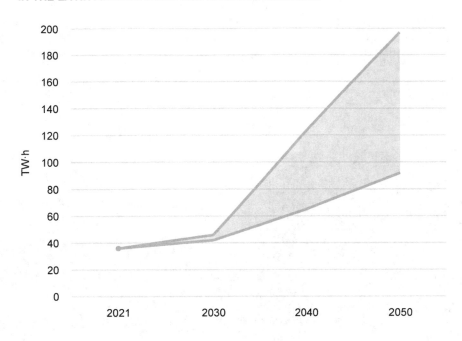

TABLE 12. TOTAL AND NUCLEAR ELECTRICAL PRODUCTION
IN THE LATIN AMERICA AND THE CARIBBEAN REGION, TW·h

Electricity Production	2021	2030		2040		2050	
		Low	High	Low	High	Low	High
Total	1 679	2 094	2 094	2 708	2 708	3 395	3 395
Nuclear	36	42	46	65	123	92	197
Nuclear as % of Electricity Production	2.1%	2.0%	2.2%	2.4%	4.5%	2.7%	5.8%

Northern, Western and Southern Europe

454

million people

Energy Overview 2021

 21.0%
of final energy consumed was electricity

 2 989TW·h
of electricity produced

 **22.5**%
of electricity produced by nuclear

FIGURE 21. FINAL ENERGY CONSUMPTION BY ENERGY SOURCE IN THE
COMBINED REGIONS OF NORTHERN, WESTERN AND SOUTHERN EUROPE

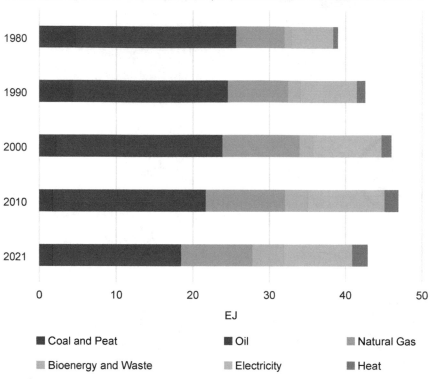

Final Energy Consumption

Since 1980 fossil fuels have continued to dominate final energy
consumption, although there has been a gradual reduction in their
combined share, which fell about 17 percentage points to about 65% in
2021.

Of all fossil fuels, oil has the largest share, although its share has been
declining over the past 40 years. In 2021 the share of oil remained
significant at about 39%.

The share of natural gas has increased by about 5 percentage points
since 1980. It accounted for over one fifth of the final energy consumed
in 2021. The share of coal has decreased by about 8 percentage points
over the past 40 years, but its share has remained at 3–4% since 2010.

In 2021 the share of electricity in final energy consumption was 21%, an
increase of almost 7 percentage points since 1980.

FIGURE 22. ELECTRICITY PRODUCTION BY ENERGY SOURCE IN THE
COMBINED REGIONS OF NORTHERN, WESTERN AND SOUTHERN EUROPE

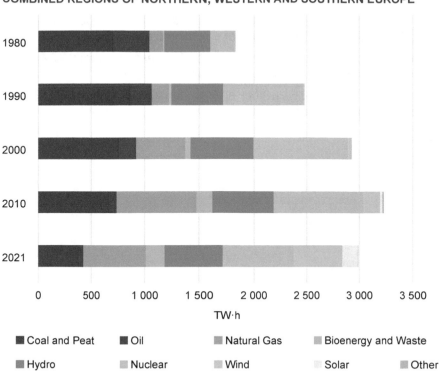

Electricity Production

In the past 40 years the combined share of fossil fuels in electricity production has effectively decreased by half. About one third of electricity was produced by fossil fuels in 2021, with natural gas being the largest contributor at about 20%. The share of natural gas has effectively tripled since 1980, whereas that of oil has declined from almost 20% in 1980 to 1.6% in 2021. Since 1980 the share of coal in electricity production has fallen from almost 40% to about 13% in 2021.

Nuclear is the largest contributor of low carbon electricity production. Its share more than doubled from 1980 to 1990 and then decreased by about 8 percentage points from 2000 to 2021 to a share of about 23%.

Over the past 40 years the share of hydro has decreased slightly by about 5 percentage points. In 2021 it was around 17%. Wind and solar did not contribute significantly to electricity production in 1980. The contributions of these energy sources have since increased substantially to a combined share of 20% in 2021.

53

Energy and Electricity Projections

- Final energy consumption is expected to decrease by about 19% by 2050, at an average annual rate of approximately 0.7%.

- Electricity consumption is projected to increase by about 27% by 2050, at an average annual rate of approximately 0.8%.

- The share of electricity in final energy consumption is expected to increase by about 12 percentage points.

FIGURE 23. FINAL CONSUMPTION OF ENERGY AND ELECTRICITY IN THE
COMBINED REGIONS OF NORTHERN, WESTERN AND SOUTHERN EUROPE

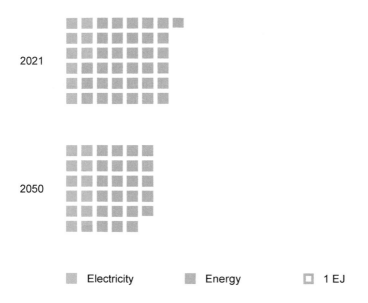

TABLE 13. FINAL CONSUMPTION OF ENERGY AND ELECTRICITY IN THE
COMBINED REGIONS OF NORTHERN, WESTERN AND SOUTHERN EUROPE, EJ

Final Consumption	2021	2030	2040	2050
Energy	42.9	41.1	37.2	34.9
Electricity	9.0	10.4	10.9	11.4
Electricity as % of Energy	*21.0%*	*25.3%*	*29.3%*	*32.7%*

Nuclear Electrical Generating Capacity Projections

- Total electrical generating capacity is projected to increase almost 11% by 2030 and by one half by 2050 compared with 2021 capacity.

- In the high case, nuclear electrical generating capacity is projected to decrease by 2030, but to a lesser extent than in the low case, and then to grow significantly by 2050, with an increase of about one third over 2021 capacity. The share of nuclear in total electrical capacity is expected to decrease about 1 percentage point by 2050.

- In the low case, nuclear electrical generating capacity is projected to decrease by almost 60% by 2050. The share of nuclear in total electrical capacity is expected to decline by almost 7 percentage points.

FIGURE 24. NUCLEAR ELECTRICAL GENERATING CAPACITY IN THE COMBINED REGIONS OF NORTHERN, WESTERN AND SOUTHERN EUROPE

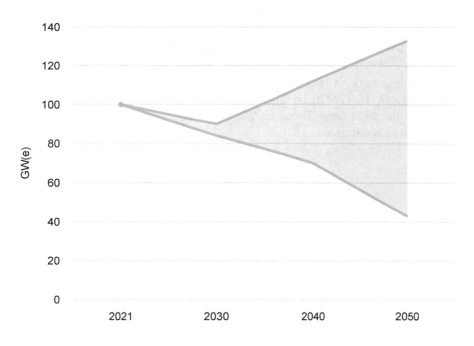

TABLE 14. TOTAL AND NUCLEAR ELECTRICAL GENERATING CAPACITY IN THE COMBINED REGIONS OF NORTHERN, WESTERN AND SOUTHERN EUROPE, GW(e)

Electrical Capacity	2021	2030		2040		2050	
		Low	High	Low	High	Low	High
Total	1 042	1 152	1 152	1 244	1 244	1 531	1 531
Nuclear	100	84	90	70	112	43	133
Nuclear as % of Electrical Capacity	*9.6%*	*7.3%*	*7.8%*	*5.6%*	*9.0%*	*2.8%*	*8.7%*

Reactor Retirements and Additions

- In the high case, it is assumed that there will be a net decrease in capacity by 2030 owing to more retirements than additions of capacity in this period. Capacity is expected to decrease by about 10 GW(e). From 2030 until 2050 a net increase of about 43 GW(e) is expected.

- In the low case, it is assumed that there will be a net decrease in capacity of 16 GW(e) by 2030, as more retirements are expected than in the high case. Between 2030 and 2050 a further reduction of 41 GW(e) is expected.

FIGURE 25. NUCLEAR CAPACITY IN THE COMBINED REGIONS OF NORTHERN, WESTERN AND SOUTHERN EUROPE: ACTUAL, RETIREMENTS AND ADDITIONS

HIGH CASE

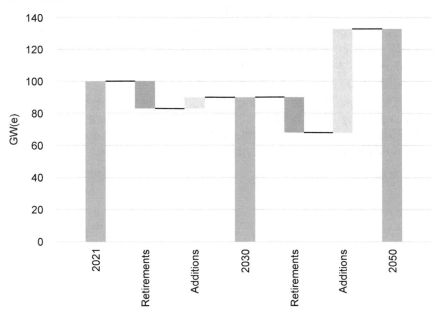

LOW CASE

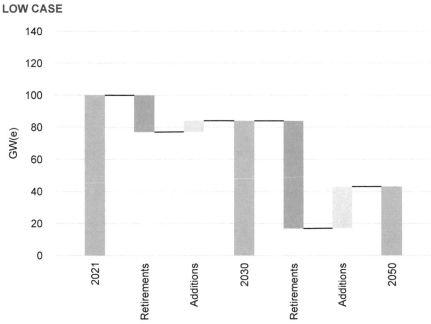

Electricity and Nuclear Production Projections

- Total electricity production is projected to increase by about 8% by 2030 and 18% by 2050, compared with 2021 production levels.

- In the high case, nuclear electricity production is projected to increase by 8% by 2030 and 62% by 2050, compared with 2021 production levels. The share of nuclear in total electricity production is expected to increase by about 8 percentage points by 2050.

- In the low case, nuclear electricity production is projected to change only slightly by 2030, but by 2050 it is expected to decrease by about 48%. The share of nuclear in total electricity production is expected to decline by more than 12 percentage points by 2050.

FIGURE 26. NUCLEAR ELECTRICITY PRODUCTION IN THE COMBINED REGIONS OF NORTHERN, WESTERN AND SOUTHERN EUROPE

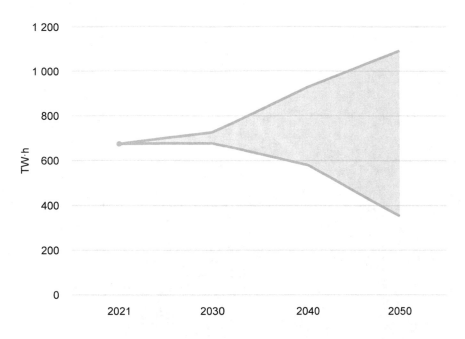

TABLE 15. TOTAL AND NUCLEAR ELECTRICAL PRODUCTION IN THE COMBINED REGIONS OF NORTHERN, WESTERN AND SOUTHERN EUROPE, TW·h

Electricity Production	2021	2030		2040		2050	
		Low	High	Low	High	Low	High
Total	2 989	3 217	3 217	3 352	3 352	3 527	3 527
Nuclear	674	678	727	580	928	353	1 092
Nuclear as % of Electricity Production	*22.5%*	*21.1%*	*22.6%*	*17.3%*	*27.7%*	*10.0%*	*31.0%*

Eastern Europe

292
million people

Energy Overview 2022

14.2%
of final energy consumed was electricity

1 591 TW·h
of electricity produced

23.9%
of electricity produced by nuclear

FIGURE 27. FINAL ENERGY CONSUMPTION BY ENERGY SOURCE
IN THE EASTERN EUROPE REGION

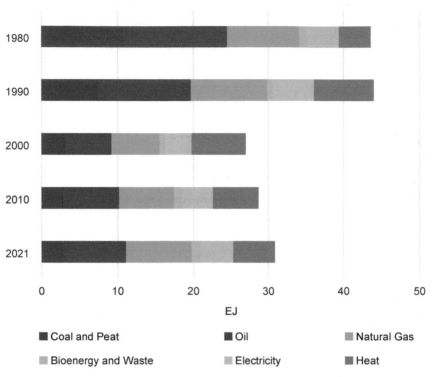

Final Energy Consumption

Since 1980 fossil fuels have accounted for the largest share of final energy consumption. Between 1980 and 2000 the combined share of fossil fuels declined by around 20 percentage points. However, from 2000 onwards there has been a gradual increase to about 64% in 2021.

Oil has the largest share of all fossil fuels, although its share has declined by about 4 percentage points over the past 40 years. In contrast, the share of natural gas has increased by about 6 percentage points since 1980. The share of coal has also declined and in 2021 was almost one third of its 1980 share.

The share of electricity has increased gradually by about 4 percentage points over the years.

With almost a doubling of its share since 1980, heat has seen the most significant change of all energy sources.

FIGURE 28. ELECTRICITY PRODUCTION BY ENERGY SOURCE
IN THE EASTERN EUROPE REGION

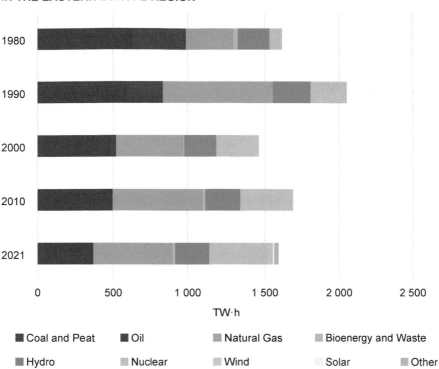

Electricity Production

Although their combined share has declined by more than 20 percentage points since 1980, fossil fuels remain the dominant sources of electricity production, with a share of about 57% in 2021.

Over the years the share of natural gas has almost doubled. In 2021 more than one third of the electricity produced was from natural gas. The share of oil has dropped significantly, from almost 23% in 1980 to less than 1% in 2021. Coal contributed about 23% of electricity production in 2021, down from 40% in 1980.

The share of nuclear has quadrupled since 1980, and nuclear accounted for 24% of the electricity produced in 2021.

The share of hydro has remained relatively stable throughout the years at about 12–14%. The combined share of solar and wind remained small at about 3% in 2021, although in 1980 these sources did not contribute to electricity production at all.

Energy and Electricity Projections

- Final consumption of energy is expected to remain stable until 2040 and then increase by only about 2.6% by 2050 compared with 2021 consumption, an average annual growth rate of approximately 0.1%.

- Electricity consumption is expected to grow at about 1.6% per year, increasing by about 57% by 2050.

- The share of electricity in final consumption of energy is expected to increase by almost 8 percentage points by 2050.

FIGURE 29. FINAL CONSUMPTION OF ENERGY AND ELECTRICITY IN THE EASTERN EUROPE REGION

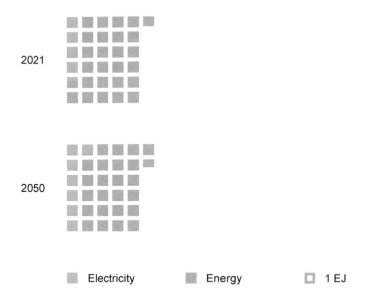

Electricity   Energy   ☐ 1 EJ

TABLE 16. FINAL CONSUMPTION OF ENERGY AND ELECTRICITY IN THE EASTERN EUROPE REGION, EJ

Final Consumption	2021	2030	2040	2050
Energy	30.9	30.8	30.9	31.7
Electricity	4.4	5.1	6.1	6.9
Electricity as % of Energy	*14.2%*	*16.6%*	*19.7%*	*21.8%*

Eastern Europe

Nuclear Electrical Generating Capacity Projections

- Total electrical generating capacity is projected to increase by about 10% by 2030 and 21% by 2050 compared with 2021 capacity.

- In the high case, nuclear electrical generating capacity is projected to almost double by 2050. However, its share of total electrical capacity is expected to increase by only about 7 percentage points.

- In the low case, nuclear electrical generating capacity is projected to remain relatively stable over the next 20 years, with a 19% increase expected by 2050. The share of nuclear in total electrical capacity is not expected to change by 2050.

FIGURE 30. NUCLEAR ELECTRICAL GENERATING CAPACITY
IN THE EASTERN EUROPE REGION

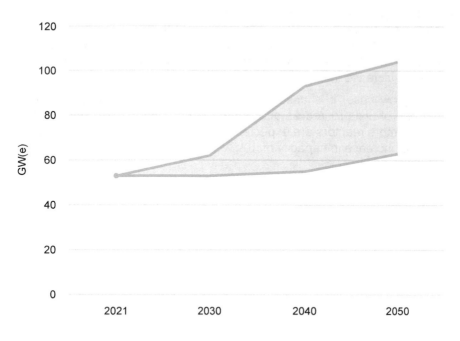

TABLE 17. TOTAL AND NUCLEAR ELECTRICAL GENERATING CAPACITY
IN THE EASTERN EUROPE REGION, GW(e)

Electrical Capacity	2021	2030		2040		2050	
		Low	High	Low	High	Low	High
Total	511	563	563	626	626	616	616
Nuclear	53	53	62	55	93	63	104
Nuclear as % of Electrical Capacity	*10.4%*	*9.4%*	*11.0%*	*8.8%*	*14.9%*	*10.2%*	*16.9%*

Reactor Retirements and Additions

- In the high case, it is assumed that twice as much capacity will be added as retired by 2030, resulting in a net increase in capacity of more than 9 GW(e). Similarly, between 2030 and 2050 almost three times as much capacity is expected to be added as retired, resulting in a net increase in capacity of almost 43 GW(e).

- In the low case, it is assumed that there will be almost the same number of units retired as added by 2030. Between 2030 and 2050 slightly more reactors are expected to be added than retired, resulting in a net increase in capacity of 10 GW(e).

FIGURE 31. NUCLEAR CAPACITY IN THE EASTERN EUROPE REGION:
ACTUAL, RETIREMENTS AND ADDITIONS

HIGH CASE

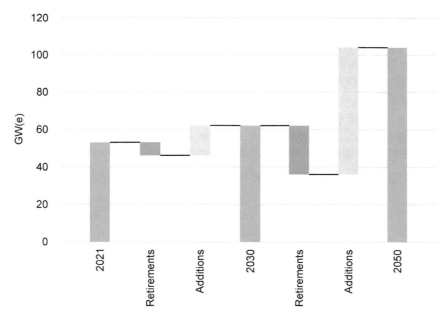

LOW CASE

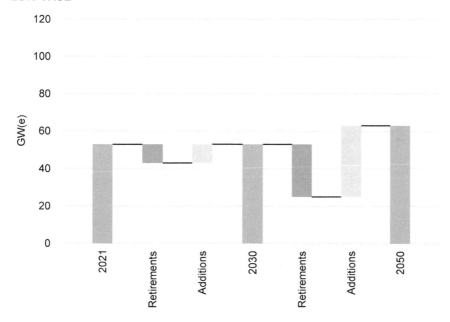

71

Electricity and Nuclear Production Projections

- Total electricity production is projected to increase by about 11% by 2030 and by about 37% by 2050 compared with 2021 production.

- In the high case, nuclear electricity production is projected to rise by 22% by 2030 compared with 2021 levels and to more than double by 2050. The share of nuclear in total electricity production is expected to increase by about 13 percentage points.

- In the low case, nuclear electricity production is projected to increase by 5% by 2030 compared with 2021 levels, and an increase of 29% is expected by 2050. The share of nuclear in total electricity production is expected to decline by less than 2 percentage points by 2050.

FIGURE 32. NUCLEAR ELECTRICITY PRODUCTION
IN THE EASTERN EUROPE REGION

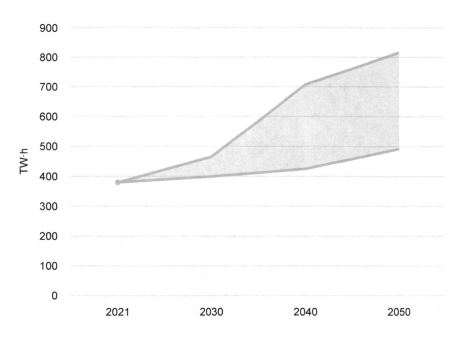

TABLE 18. TOTAL AND NUCLEAR ELECTRICAL PRODUCTION
IN THE EASTERN EUROPE REGION, TW·h

Electricity Production	2021	2030		2040		2050	
		Low	High	Low	High	Low	High
Total	1 591	1 763	1 763	1 984	1 984	2 186	2 186
Nuclear	380	399	465	425	707	492	816
Nuclear as % of Electricity Production	*23.9%*	*22.6%*	*26.4%*	*21.4%*	*35.6%*	*22.5%*	*37.3%*

Africa

1 377
million people

Energy Overview 2021

 9.8%
of final energy consumed was electricity

 852TW·h
of electricity produced

 **1.4%**
of electricity produced by nuclear

Africa

FIGURE 33. FINAL ENERGY CONSUMPTION BY ENERGY SOURCE
IN THE AFRICA REGION

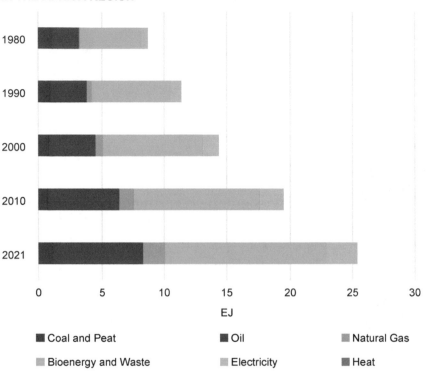

Final Energy Consumption

Bioenergy and waste has accounted for the largest share of final energy consumption over the past 40 years at around 50%.

The combined share of fossil fuels has been relatively stable since 1980 at about 40%. The share of natural gas has increased by about 6 percentage points over the past 40 years, while the share of oil has increased by about 3 percentage points. The share of coal has gradually decreased by about 7 percentage points.

The share of electricity has increased a few percentage points since 1980 to reach almost 10% in 2021.

FIGURE 34. ELECTRICITY PRODUCTION BY ENERGY SOURCE IN
THE AFRICA REGION

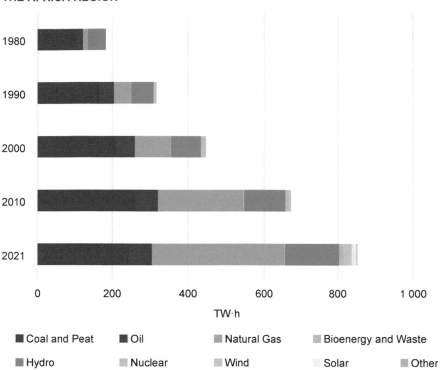

Electricity Production

From 1980 until 2010 the combined share of fossil fuels in electricity production gradually increased by about 6 percentage points. From 2010 to 2021 their combined share decreased by the same amount but was still 77% of electricity production in 2021.

Since 1980 the share of natural gas has steadily increased by about 30 percentage points, whereas the share of coal has declined almost 50%. The share of oil has decreased by about 5 percentage points.

The share of nuclear was around 2–3% from 1990 to 2010 and 1.4% in 2021.

Hydro was the largest contributor of low carbon energy, accounting for more than 17% of electricity production in 2021, although its share has decreased by about 10 percentage points over the past 40 years. The share of wind and solar has increased slightly since 2000, rising from less than 1% to about 4% in 2021.

Energy and Electricity Projections

- Final energy consumption is expected to increase 16% from 2021 levels by 2030 and by 76% by 2050, at an average annual rate of approximately 2%.

- Electricity consumption will grow much faster, at an average annual rate of approximately 5%, and is expected to increase more than fourfold from 2021 levels by 2050.

- Over the next 29 years the share of electricity in final energy consumption is expected to more than double from its 2021 share.

FIGURE 35. FINAL CONSUMPTION OF ENERGY AND ELECTRICITY
IN THE AFRICA REGION

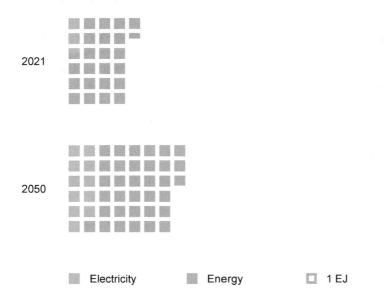

TABLE 19. FINAL CONSUMPTION OF ENERGY AND ELECTRICITY
IN THE AFRICA REGION, EJ

Final Consumption	2021	2030	2040	2050
Energy	25.5	29.5	35.7	44.9
Electricity	2.5	3.9	6.7	10.9
Electricity as % of Energy	*9.8%*	*13.2%*	*18.8%*	*24.3%*

Per Capita Energy and Electricity

- Only about 25% of the electricity produced in Africa is consumed by the residential sector.

- Electricity consumption on a per capita basis is expected to more than double from 0.5 MW·h per person in 2021 to 1.2 MW·h per person in 2050. This would be enough electricity to power one high efficiency modern (circa 2020) washing machine or one small high efficiency (induction) electric stove for 30 minutes per day.

- In 2010 the world average electricity consumption for households with electricity access was about 3.5 MW·h, almost six times that for the residential sector in Africa in 2021.

FIGURE 36. PER CAPITA ELECTRICITY CONSUMPTION
IN THE AFRICA REGION

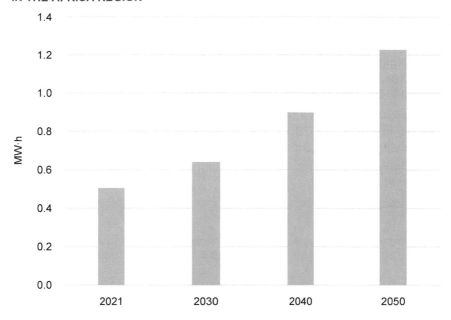

FIGURE 37. PER CAPITA FINAL ENERGY CONSUMPTION
IN THE AFRICA REGION

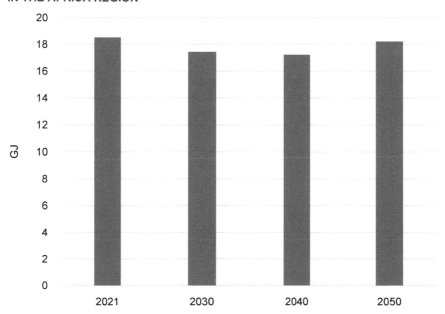

Nuclear Electrical Generating Capacity Projections

- Total electrical generating capacity is expected to increase by 51% by 2030 and to undergo a fourfold increase by 2050.

- In the high case, nuclear electrical generating capacity is projected to increase by 58% by 2030 and to undergo more than an 11-fold increase by 2050 compared with 2021 capacity.

- In the low case, nuclear electrical generating capacity is projected to remain constant to 2030, and by 2050 it is expected to undergo close to a fivefold increase compared with 2021 levels.

FIGURE 38. NUCLEAR ELECTRICAL GENERATING CAPACITY
IN THE AFRICA REGION

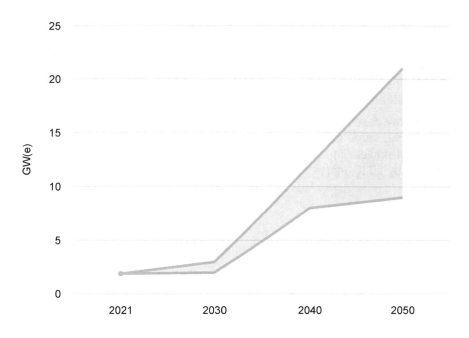

TABLE 20. TOTAL AND NUCLEAR ELECTRICAL GENERATING CAPACITY
IN THE AFRICA REGION, GW(e)

Electrical Capacity	2021	2030		2040		2050	
		Low	High	Low	High	Low	High
Total	250	378	378	617	617	1 028	1 028
Nuclear	1.9	2	3	8	12	9	21
Nuclear as % of Electrical Capacity	*0.8%*	*0.5%*	*0.8%*	*1.3%*	*1.9%*	*0.9%*	*2.0%*

Electricity and Nuclear Production Projections

- Total electricity production is projected to increase by 51% by 2030 and to increase more than fourfold by 2050.

- In the high case, nuclear electricity production is expected to almost double from 2021 levels by 2030 and to increase more than 12-fold by 2050. The share of nuclear in total electricity production is expected to more than triple.

- In the low case, nuclear electricity production is expected to remain almost the same to 2030 and to increase about sixfold by 2050. The share of nuclear in total electricity production is expected to decline slightly by 2030, increasing again thereafter to reach 2% by 2050.

FIGURE 39. NUCLEAR ELECTRICITY PRODUCTION IN THE AFRICA REGION

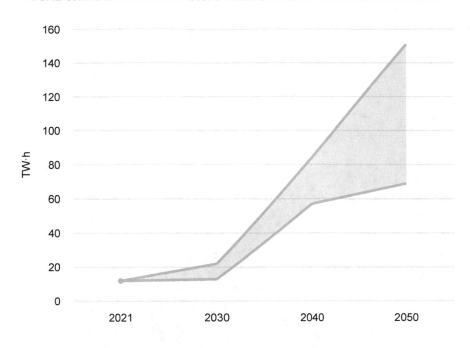

TABLE 21. TOTAL AND NUCLEAR ELECTRICAL PRODUCTION IN THE AFRICA REGION, TW·h

Electricity Production	2021	2030 Low	2030 High	2040 Low	2040 High	2050 Low	2050 High
Total	852	1 285	1 285	2 205	2 205	3 533	3 533
Nuclear	12	13	22	57	84	69	151
Nuclear as % of Electricity Production	*1.4%*	*1.0%*	*1.7%*	*2.6%*	*3.8%*	*2.0%*	*4.3%*

Western Asia

288
million people

Energy Overview 2021

 19.7%
of final energy consumed was electricity

 1 266TW·h
of electricity produced

 **0.9**%
of electricity produced by nuclear

FIGURE 40. FINAL ENERGY CONSUMPTION BY ENERGY SOURCE
IN THE WESTERN ASIA REGION

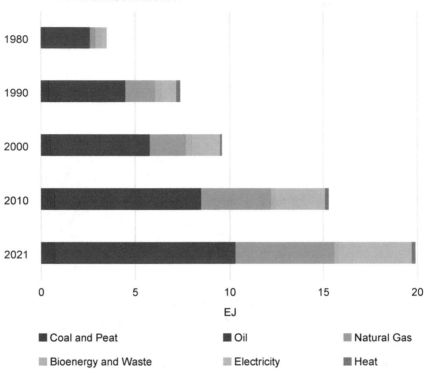

Final Energy Consumption

Fossil fuels have continued to dominate final energy consumption, with a stable share of about 80% since 1980.

At about 50% in 2021, oil had the largest share of final energy of all fossil fuels, despite a reduction of 19 percentage points since 1980. The share of natural gas has increased steadily since 1980 and was the second largest, accounting for more than a quarter of final energy consumption in 2021. The share of coal was less than 4% in 2021, remaining relatively small and decreasing by a few percentage points since 1980.

At about 20% in 2021, the share of electricity in final energy consumption has more than doubled since 1980.

The share of bioenergy and waste in final energy consumption has declined gradually over the years, decreasing from about 9% in 1980 to less than 1% in 2021.

FIGURE 41. ELECTRICITY PRODUCTION BY ENERGY SOURCE
IN THE WESTERN ASIA REGION

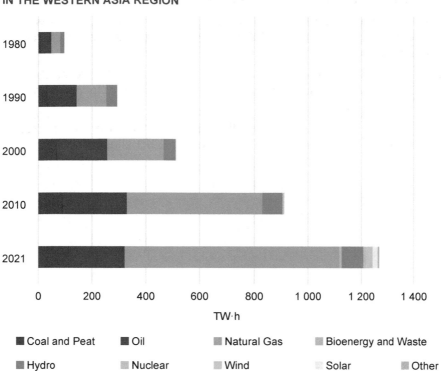

TW·h

■ Coal and Peat   ■ Oil   ■ Natural Gas   ■ Bioenergy and Waste

■ Hydro   ■ Nuclear   ■ Wind   ☐ Solar   ■ Other

Electricity Production

With a share of almost 90%, fossil fuels — particularly natural gas — have remained dominant sources of electricity production since 1980.

Hydro remains the largest contributor of low carbon electricity, accounting for about 6% of total production, although its share has declined by more than half since 1980.

The share of nuclear in electricity production remains small at about 1%.

In recent years, solar and wind have begun generating electricity, and in 2021 their combined share was more than 3%.

Energy and Electricity Projections

- Final energy consumption is expected to increase from 2021 levels by about 9% by 2030 and about 26% by 2050, at an average annual rate of approximately 0.8%.

- Electricity consumption is expected to grow at a faster rate of about 2.8% per year. Electricity consumption is expected to more than double by 2050.

- By 2050 the share of electricity in final energy consumption is expected to increase by 15 percentage points from its 2021 share.

FIGURE 42. FINAL CONSUMPTION OF ENERGY AND ELECTRICITY IN THE WESTERN ASIA REGION

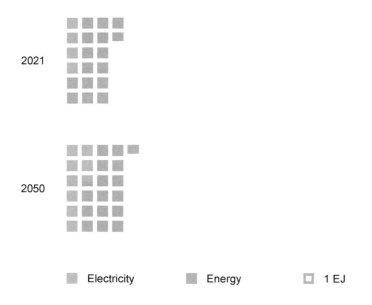

■ Electricity	■ Energy	☐ 1 EJ

TABLE 22. FINAL CONSUMPTION OF ENERGY AND ELECTRICITY IN THE WESTERN ASIA REGION, EJ

Final Consumption	2021	2030	2040	2050
Energy	19.8	21.5	22.7	24.8
Electricity	3.9	5.3	6.9	8.6
Electricity as % of Energy	*19.7%*	*24.7%*	*30.4%*	*34.7%*

Nuclear Electrical Generating Capacity Projections

- Total electrical generating capacity is expected to increase by about 32% by 2030 and about 160% by 2050.

- In the high case, nuclear electrical generating capacity is projected to undergo an almost threefold increase by 2030 and about an eightfold increase by 2050 compared with 2021 capacity.

- In the low case, nuclear electrical generating capacity is projected to increase almost threefold by 2030 and more than a fourfold by 2050 compared with 2021 capacity.

- The share of nuclear in total electrical generating capacity is expected to increase by 2050 in both the high and the low case.

FIGURE 43. NUCLEAR ELECTRICAL GENERATING CAPACITY
IN THE WESTERN ASIA REGION

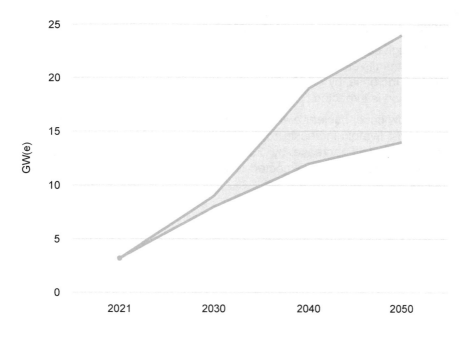

TABLE 23. TOTAL AND NUCLEAR ELECTRICAL GENERATING CAPACITY
IN THE WESTERN ASIA REGION, GW(e)

Electrical Capacity	2021	2030		2040		2050	
		Low	High	Low	High	Low	High
Total	399	525	525	716	716	1 023	1 023
Nuclear	3.2	8	9	12	19	14	24
Nuclear as % of Electrical Capacity	*0.8%*	*1.5%*	*1.7%*	*1.7%*	*2.7%*	*1.4%*	*2.3%*

Electricity and Nuclear Production Projections

- Total electricity production is projected to increase by about 36% by 2030 and to more than double by 2050.

- In the high case, nuclear electricity production is expected to undergo a more than fivefold increase from 2021 levels by 2030 and an almost 16-fold increase by 2050. The share of nuclear in total electricity production is expected to increase by about 6 percentage points.

- In the low case, nuclear electricity production is expected to undergo an almost fivefold increase from 2021 levels by 2030, rising to more than a ninefold increase by 2050. The share of nuclear in total electricity production is expected to increase by about 3 percentage points.

FIGURE 44. NUCLEAR ELECTRICITY PRODUCTION
IN THE WESTERN ASIA REGION

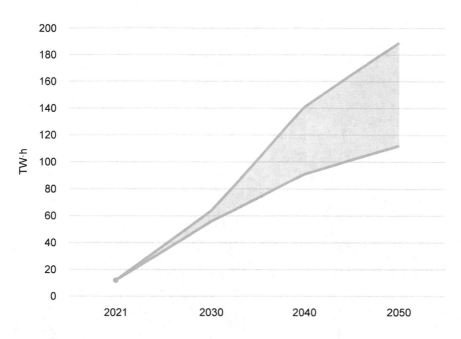

TABLE 24. TOTAL AND NUCLEAR ELECTRICAL PRODUCTION
IN THE WESTERN ASIA REGION, TW·h

Electricity Production	2021	2030		2040		2050	
		Low	High	Low	High	Low	High
Total	1 266	1 717	1 717	2 256	2 256	2 817	2 817
Nuclear	12	56	64	91	141	112	189
Nuclear as % of Electricity Production	*0.9%*	*3.3%*	*3.7%*	*4.0%*	*6.2%*	*4.0%*	*6.7%*

Southern Asia

1 980

million people

FIGURE 46. ELECTRICITY PRODUCTION BY ENERGY SOURCE
IN THE SOUTHERN ASIA REGION

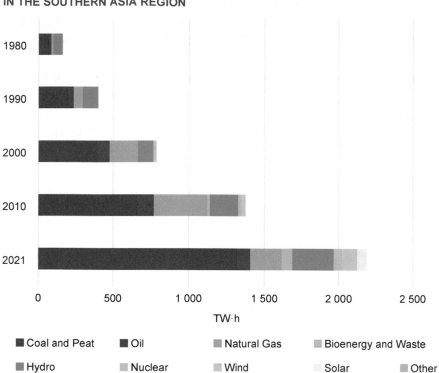

TW·h

■ Coal and Peat   ■ Oil   ■ Natural Gas   ■ Bioenergy and Waste

■ Hydro   ■ Nuclear   ■ Wind   ■ Solar   ■ Other

Electricity Production

From 1980 to 2000, fossil fuels increased their share of electricity production from about 60% to more than 80%. Their share has since declined to about 74% in 2021.

Coal is the largest source of electricity with a share of 60%, an increase of about 24 percentage points since 1980. The share of natural gas has increased since 1980, accounting for about 10% of the electricity produced in 2021. The share of oil has decreased 9 percentage points since 1980 to about 4% in 2021.

Hydro remains the largest contributor of low carbon electricity, accounting for 13% of total production, although its share has decreased by about 26 percentage points since 1980. In recent years, the share of solar and wind has undergone a rapid increase, rising from less than 1% in 2000 to more than 7% in 2021.

The share of nuclear was about 3% in 2021.

Energy and Electricity Projections

- Final energy consumption is expected to increase by about 40% from 2021 levels by 2030 and to double by 2050, at an average annual rate of approximately 2.5%.

- Electricity consumption is expected to grow at a faster rate of 4.4% per year. Electricity consumption is expected to more than triple by 2050.

- By 2050 the share of electricity in final energy consumption is expected to increase by 10 percentage points from its 2021 share.

FIGURE 47. FINAL CONSUMPTION OF ENERGY AND ELECTRICITY
IN THE SOUTHERN ASIA REGION

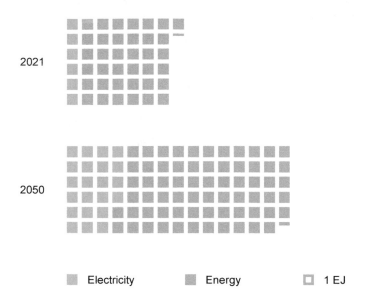

TABLE 25. FINAL CONSUMPTION OF ENERGY AND ELECTRICITY
IN THE SOUTHERN ASIA REGION, EJ

Final Consumption	2021	2030	2040	2050
Energy	43.2	60.6	74.7	89.3
Electricity	6.4	10.7	16.1	22.1
Electricity as % of Energy	*14.8%*	*17.7%*	*21.6%*	*24.7%*

Nuclear Electrical Generating Capacity Projections

- Total electrical generating capacity is expected to nearly double by 2030 and to increase fivefold by 2050.

- In the high case, nuclear electrical generating capacity is projected to triple by 2030 and to undergo an eightfold increase by 2050 compared with 2021 capacity. The share of nuclear in total electrical generating capacity is expected to increase by about 1 percentage point by 2050.

- In the low case, nuclear electrical generating capacity is projected to almost double by 2030 and increase fivefold by 2050. The share of nuclear in total electrical generating capacity is expected to remain roughly at the 2021 level.

FIGURE 48. NUCLEAR ELECTRICAL GENERATING CAPACITY
IN THE SOUTHERN ASIA REGION

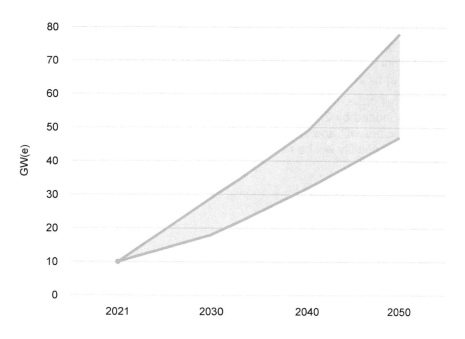

TABLE 26. TOTAL AND NUCLEAR ELECTRICAL GENERATING CAPACITY
IN THE SOUTHERN ASIA REGION, GW(e)

Electrical Capacity	2021	2030		2040		2050	
		Low	High	Low	High	Low	High
Total	597	1 127	1 127	1 806	1 806	3 014	3 014
Nuclear	10	18	29	32	49	47	78
Nuclear as % of Electrical Capacity	*1.7%*	*1.6%*	*2.6%*	*1.8%*	*2.7%*	*1.6%*	*2.6%*

Reactor Retirements and Additions

- In the high case, it is assumed that only about 6% of the 2021 nuclear electrical generating capacity will be retired by 2030 and that 20% of the 2021 nuclear electrical generating capacity will be retired by 2050. This is expected to contribute to net capacity additions of about 19 GW(e) by 2030 and 49 GW(e) over the subsequent 20 years.

- In the low case, it is assumed there will be about 8 GW(e) of net capacity added by 2030. Between 2030 and 2050 it is expected that new reactors will add about 30 GW(e) of capacity and only a few GW(e) of capacity will be retired.

FIGURE 49. NUCLEAR CAPACITY IN THE SOUTHERN ASIA REGION:
ACTUAL, RETIREMENTS AND ADDITIONS

HIGH CASE

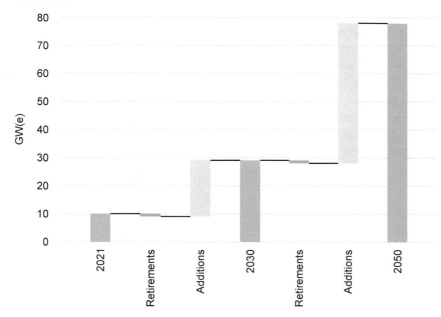

LOW CASE

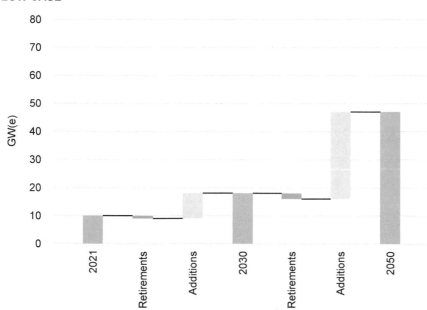

Electricity and Nuclear Production Projections

- Total electrical production is projected to increase by 63% by 2030 and by more than threefold by 2050 compared with 2021 production.

- In the high case, nuclear electricity production is expected to more than triple from 2021 levels by 2030 and to increase about tenfold by 2050. The share of nuclear in total electricity production is expected to increase by about 6 percentage points.

- In the low case, nuclear electricity production is expected to more than double from 2021 levels by 2030 and to increase more than sixfold by 2050. The share of nuclear in total electricity production is expected to nearly double.

**FIGURE 50. NUCLEAR ELECTRICITY PRODUCTION
IN THE SOUTHERN ASIA REGION**

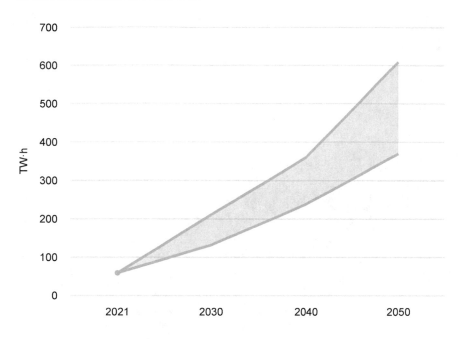

**TABLE 27. TOTAL AND NUCLEAR ELECTRICAL PRODUCTION
IN THE SOUTHERN ASIA REGION, TW·h**

Electricity Production	2021	2030		2040		2050	
		Low	High	Low	High	Low	High
Total	2 188	3 556	3 556	5 303	5 303	7 253	7 253
Nuclear	59	132	211	239	361	370	609
Nuclear as % of Electricity Production	*2.7%*	*3.7%*	*5.9%*	*4.5%*	*6.8%*	*5.1%*	*8.4%*

Central and Eastern Asia

1 739

million people

Energy Overview 2021

24.5%
of final energy consumed was electricity

10 208 TW·h
of electricity produced

6.1%
of electricity produced by nuclear

FIGURE 51. FINAL ENERGY CONSUMPTION BY ENERGY SOURCE
IN THE COMBINED REGIONS OF CENTRAL AND EASTERN ASIA

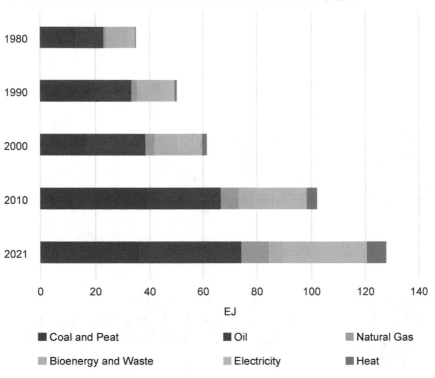

Final Energy Consumption

Since 1980 fossil fuels have dominated final energy consumption with a combined share that has remained around 65–70%.

The share of natural gas has increased fourfold since 1980, whereas oil has maintained a relatively consistent share of about 30%. The share of coal was almost 28% in 2021.

The share of electricity has more than doubled since 1980, accounting for almost a quarter of final energy consumption in 2021.

The share of bioenergy and waste in final energy consumption has decreased by about 18 percentage points since 1980.

The share of heat has increased from less than 1% in 1980 to almost 6% in 2021.

FIGURE 52. ELECTRICITY PRODUCTION BY ENERGY SOURCE
IN THE COMBINED REGIONS OF CENTRAL AND EASTERN ASIA

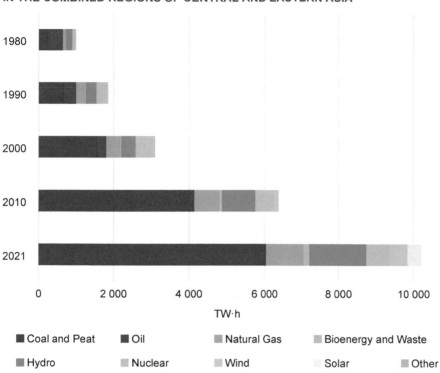

Electricity Production

With a share of about 70%, fossil fuels — particularly coal — have remained dominant sources of electricity production since 1980.

The share of coal has increased more than 35 percentage points since 1980. The share of natural gas increased from 1980 to 1990 and then declined to about 10%. Of all fossil fuels, the share of oil has experienced the most significant change, decreasing from about 42% in 1980 to below 1% in 2021.

Hydro was the largest contributor of low carbon electricity, accounting for 15% of total production in 2021. Its share has remained relatively stable over the past 40 years. In recent years, the share of solar and wind has increased rapidly, rising from less than 1% in 2010 to more than 8% in 2021.

The share of nuclear increased between 1980 and 2000 but has since declined, falling to about 6% in 2021.

Energy and Electricity Projections

- Final energy consumption is expected to increase by about 10% from 2021 levels by 2030 and by about 15% by 2050, at an average annual rate of approximately 0.5%.

- Electricity consumption is expected to grow at a faster rate of about 2% per year. Electricity consumption is expected to increase by about 74% from 2021 levels by 2050.

- By 2050 the share of electricity in final energy consumption is expected to increase by about 12 percentage points from its 2021 share.

FIGURE 53. FINAL CONSUMPTION OF ENERGY AND ELECTRICITY
IN THE COMBINED REGIONS OF CENTRAL AND EASTERN ASIA

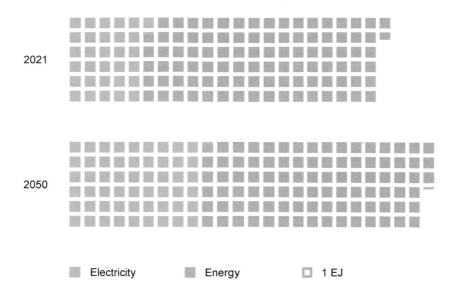

TABLE 28. FINAL CONSUMPTION OF ENERGY AND ELECTRICITY
IN THE COMBINED REGIONS OF CENTRAL AND EASTERN ASIA, EJ

Final Consumption	2021	2030	2040	2050
Energy	127.7	140.2	144.7	147.2
Electricity	31.3	39.6	47.9	54.3
Electricity as % of Energy	*24.5%*	*28.2%*	*33.1%*	*36.9%*

Nuclear Electrical Generating Capacity Projections

- Total electrical generating capacity is expected to increase by about 20% by 2030 and by more than 70% by 2050.

- In the high case, nuclear electrical generating capacity is projected to increase by about 60% by 2030 and to more than triple by 2050 compared with 2021 capacity. The share of nuclear in total electrical generating capacity is expected to increase by about 3 percentage points by 2050.

- In the low case, nuclear electrical generating capacity is projected to increase by less than 15% by 2030 and by about 60% by 2050 compared with 2021 capacity. The share of nuclear in total electrical generating capacity is expected to decrease slightly by 2050.

FIGURE 54. NUCLEAR ELECTRICAL GENERATING CAPACITY
IN THE COMBINED REGIONS OF CENTRAL AND EASTERN ASIA

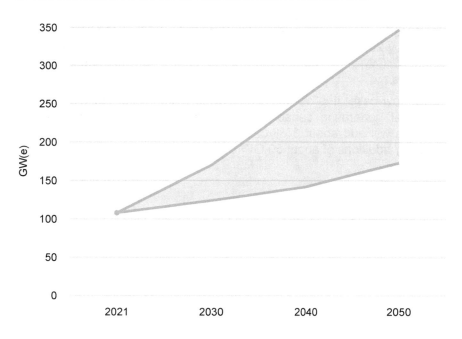

TABLE 29. TOTAL AND NUCLEAR ELECTRICAL GENERATING CAPACITY
IN THE COMBINED REGIONS OF CENTRAL AND EASTERN ASIA, GW(e)

Electrical Capacity	2021	2030		2040		2050	
		Low	High	Low	High	Low	High
Total	3 077	3 678	3 678	4 671	4 671	5 324	5 324
Nuclear	108	124	170	142	260	173	347
Nuclear as % of Electrical Capacity	*3.5%*	*3.4%*	*4.6%*	*3.0%*	*5.6%*	*3.2%*	*6.5%*

Reactor Retirements and Additions

- In the high case, it is assumed that about 2% of the 2021 nuclear electrical generating capacity will be retired by 2030 and 15% will be retired by 2050. This is expected to result in net capacity additions of about 62 GW(e) by 2030 and about 177 GW(e) over the subsequent 20 years.

- In the low case, it is assumed that about 14% of the 2021 nuclear electrical generating capacity will be retired by 2030, while new reactors will add about 30% capacity. Between 2030 and 2050 net capacity additions of about 50 GW(e) are expected.

FIGURE 55. NUCLEAR CAPACITY IN THE COMBINED REGIONS OF
CENTRAL AND EASTERN ASIA: ACTUAL, RETIREMENTS AND ADDITIONS

HIGH CASE

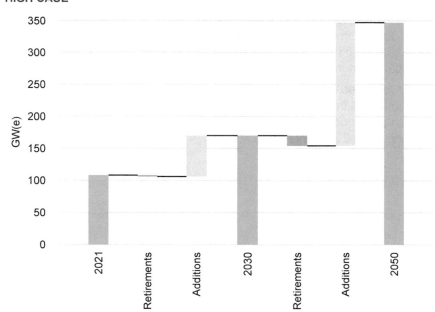

LOW CASE

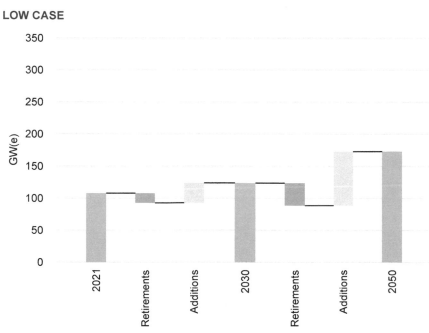

Electricity and Nuclear Production Projections

- Total electricity production is projected to increase by 62% by 2050.

- In the high case, nuclear electricity production is expected to more than double from 2021 levels by 2030 and to undergo a 4.5-fold increase by 2050. The share of nuclear in total electricity production is expected to increase by about 11 percentage points.

- In the low case, nuclear electricity production is expected to increase by about 50% from 2021 levels by 2030 and by almost 160% by 2050. The share of nuclear in total electricity production is expected increase by more than 3 percentage points.

FIGURE 56. NUCLEAR ELECTRICITY PRODUCTION
IN THE COMBINED REGIONS OF CENTRAL AND EASTERN ASIA

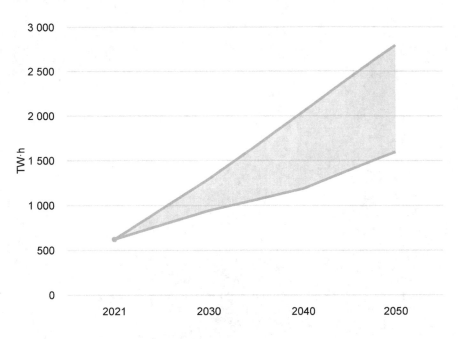

TABLE 30. TOTAL AND NUCLEAR ELECTRICAL PRODUCTION
IN THE COMBINED REGIONS OF CENTRAL AND EASTERN ASIA, TW·h

Electricity Production	2021	2030		2040		2050	
		Low	High	Low	High	Low	High
Total	10 208	12 101	12 101	14 632	14 632	16 576	16 576
Nuclear	622	943	1 293	1 188	2 047	1 597	2 793
Nuclear as % of Electricity Production	6.1%	7.8%	10.7%	8.1%	14.0%	9.6%	16.8%

South-eastern Asia

673
million people

Energy Overview 2021

 18.3%
of final energy consumed was electricity

 1 132TW·h
of electricity produced

 **0**%
of electricity produced by nuclear

FIGURE 57. FINAL ENERGY CONSUMPTION BY ENERGY SOURCE
IN THE SOUTH-EASTERN ASIA REGION

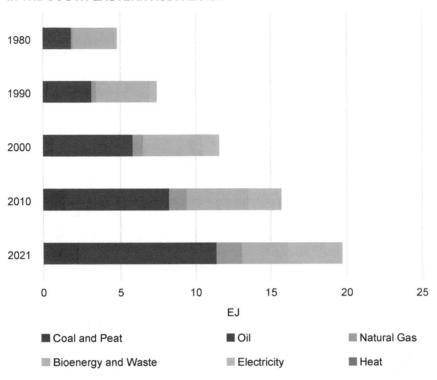

Final Energy Consumption

From 1980 to 1990, bioenergy and waste accounted for the largest share of final energy consumption.

Since 2000, fossil fuels have dominated final energy consumption, with oil having the largest share at about 47%. The share of coal has gradually increased over the past 40 years, reaching 11% in 2021, an increase of 9 percentage points. The share of natural gas has quadrupled since 1980, reaching almost 9% in 2020.

At 18% in 2021, electricity's share has increased fourfold since 1980.

FIGURE 58. ELECTRICITY PRODUCTION BY ENERGY SOURCE
IN THE SOUTH-EASTERN ASIA REGION

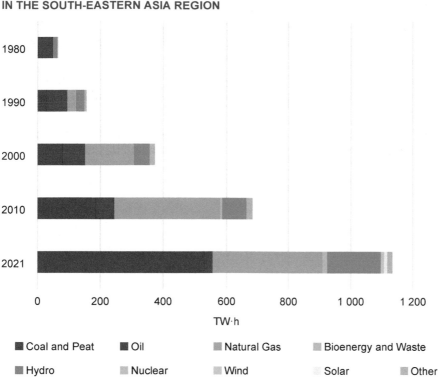

Electricity Production

With a share of about 80% over the past 40 years, fossil fuels have remained dominant sources of electricity production.

The share of coal has increased 35 percentage points since 1980 and reached almost 47% in 2021, whereas oil's share has declined by some 70 percentage points to about 2% in 2021. The share of natural gas has increased about 30 percentage points since 1980.

Hydro remains the largest contributor of low carbon electricity, accounting for 15% of total electricity production in 2021. The share of 'other' sources (mainly geothermal) increased by about 1 percentage point between 1980 and 2000, but has since fallen, reaching about 1.5% in 2021. Solar and wind have recently begun contributing to electricity generation, accounting for slightly below 2% in 2021.

Energy and Electricity Projections

- Final energy consumption is expected to increase by about 33% from 2021 levels by 2030 and by 75% by 2050, at an average annual rate of approximately 2%.

- Electricity consumption is expected to grow at a faster rate of 4.1% per year. Electricity consumption is expected to more than triple by 2050.

- By 2050 the share of electricity in final energy consumption is expected to increase by about 15 percentage points from its 2021 share.

FIGURE 59. FINAL CONSUMPTION OF ENERGY AND ELECTRICITY
IN THE SOUTH-EASTERN ASIA REGION

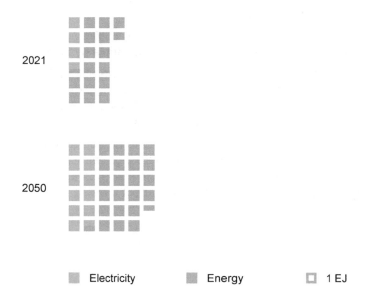

TABLE 31. FINAL CONSUMPTION OF ENERGY AND ELECTRICITY
IN THE SOUTH-EASTERN ASIA REGION, EJ

Final Consumption	2021	2030	2040	2050
Energy	19.7	26.2	29.0	34.5
Electricity	3.6	6.0	8.5	11.5
Electricity as % of Energy	*18.3%*	*22.9%*	*29.3%*	*33.3%*

Nuclear Electrical Generating Capacity Projections

- Total electrical generating capacity is expected to increase by about 45% by 2030 and to almost triple by 2050.

- Total electricity production is projected to increase by 66% by 2030 compared with 2021 production levels and to more than triple by 2050.

- In the high case, nuclear reactors are projected to be operational by 2040 and by 2050 nuclear electrical generating capacity is expected to triple compared with 2040 capacity. The share of nuclear in total electricity production is expected to reach about 1.6%.

- In the low case, nuclear reactors are also projected to be operational by 2040 and by 2050 nuclear electrical generating capacity is expected to triple compared with 2040 capacity. The share of nuclear in total electricity production is expected to reach about 0.3%.

TABLE 32. TOTAL AND NUCLEAR ELECTRICAL GENERATING
CAPACITY IN THE SOUTH-EASTERN ASIA REGION, GW(e)

Electrical Capacity	2021	2030		2040		2050	
		Low	High	Low	High	Low	High
Total	315	458	458	671	671	928	928
Nuclear	0.0	0	0	1	5	3	15
Nuclear as % of Electrical Capacity	0.0%	0.0%	0.0%	0.1%	0.7%	0.3%	1.6%

TABLE 33. TOTAL AND NUCLEAR ELECTRICAL PRODUCTION
IN THE SOUTH-EASTERN ASIA REGION, TW·h

Electricity Production	2021	2030		2040		2050	
		Low	High	Low	High	Low	High
Total	1 132	1 880	1 880	2 612	2 612	3 496	3 496
Nuclear	0	0	0	8	37	24	118
Nuclear as % of Electricity Production	0.0%	0.0%	0.0%	0.3%	1.4%	0.7%	3.4%

127

Oceania

44

million people

Energy Overview 2021

22.0%
of final energy consumed was electricity

302TW·h
of electricity produced

0%
of electricity produced by nuclear

FIGURE 60. FINAL ENERGY CONSUMPTION BY ENERGY SOURCE
IN THE OCEANIA REGION

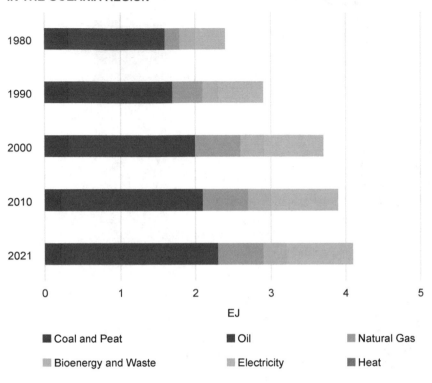

Legend:
- Coal and Peat
- Oil
- Natural Gas
- Bioenergy and Waste
- Electricity
- Heat

Final Energy Consumption

Since 1980 fossil fuels have continued to dominate final energy
consumption, although there was a gradual reduction in their combined
share from 76% in 1980 to 69% in 2010. Between 2010 and 2021 their
combined share increased by 1 percentage point.

Oil has the largest share of all the fossil fuels, having remained at about
50% since 1990. The share of natural gas has remained at about 15%
since 1990, whereas coal's share has declined by almost 5 percentage
points during the same period.

The share of electricity is more than one fifth of final energy consumption,
an increase of 8 percentage points since 1980.

FIGURE 61. ELECTRICITY PRODUCTION BY ENERGY SOURCE
IN THE OCEANIA REGION

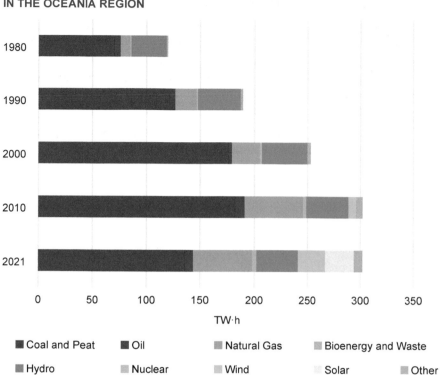

Electricity Production

With a share of more than 70%, fossil fuels — particularly coal — have remained dominant sources of electricity production over the past 40 years. From 1980 to 2010 their share increased some 10 percentage points and then declined by 15 percentage points by 2021.

The share of natural gas has more than doubled since 1980, whereas oil's share has fallen by half. The share of coal increased from almost 60% in 1980 to almost 70% by 2000, and then fell to about 45% by 2021.

The share of hydro has declined by more than half since 1980, reaching about 13% in 2021. The combined share of solar and wind has increased from 0.1% in 2000 to about 17% in 2021.

Energy and Electricity Projections

- Final energy consumption is expected to increase by 12% from 2021 levels by 2030 and by almost 15% by 2050, at an average annual rate of approximately 0.5%.

- Electricity consumption is expected to grow at a faster rate of about 1.3% per year. Electricity consumption is expected to increase by about 44% by 2050.

- By 2050 the share of electricity in final energy consumption is expected to increase by about 6 percentage points from its 2021 share.

FIGURE 62. FINAL CONSUMPTION OF ENERGY AND ELECTRICITY
IN THE OCEANIA REGION

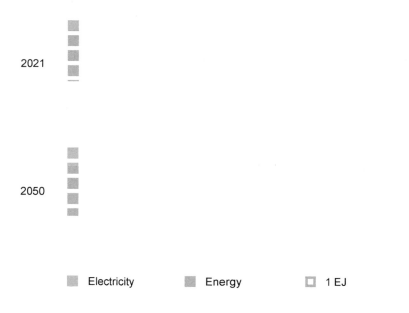

2021

2050

Electricity   Energy   1 EJ

TABLE 34. FINAL CONSUMPTION OF ENERGY AND ELECTRICITY
IN THE OCEANIA REGION, EJ

Final Consumption	2021	2030	2040	2050
Energy	4.1	4.6	4.6	4.7
Electricity	0.9	1.1	1.2	1.3
Electricity as % of Energy	*22.0%*	*23.9%*	*26.1%*	*27.7%*

Nuclear Electrical Generating Capacity Projections

- Total electrical generating capacity is expected to increase from 2021 levels by about 20% by 2030 and by about 90% by 2050.

- Total electricity production is projected to increase by about 9% by 2030 and by about 34% by 2050 compared with 2021 production levels.

- In the high case, nuclear power is projected to generate electricity by the middle of the century. The share of nuclear in total electrical generating capacity is expected to reach about 1%.

- In the low case, nuclear power is not projected to be introduced into the electricity generation system.

TABLE 35. TOTAL AND NUCLEAR ELECTRICAL GENERATING CAPACITY
IN THE OCEANIA REGION, GW(e)

Electrical Capacity	2021	2030		2040		2050	
		Low	High	Low	High	Low	High
Total	94	114	114	131	131	177	177
Nuclear	0.0	0	0	0	0	0	2
Nuclear as % of Electrical Capacity	0.0%	0.0%	0.0%	0.0%	0.0%	0.0%	1.1%

TABLE 36. TOTAL AND NUCLEAR ELECTRICAL PRODUCTION
IN THE OCEANIA REGION, TW·h

Electricity Production	2021	2030		2040		2050	
		Low	High	Low	High	Low	High
Total	302	328	328	365	365	405	405
Nuclear	0	0	0	0	0	0	14
Nuclear as % of Electricity Production	0.0%	0.0%	0.0%	0.0%	0.0%	0.0%	3.5%

135

REFERENCES

[1] INTERNATIONAL ENERGY AGENCY, World Energy Outlook 2021, IEA, Paris (2021), https://www.iea.org/reports/world-energy-outlook-2021

[2] UNITED STATES ENERGY INFORMATION ADMINISTRATION, International Energy Outlook 2021, U.S. Department of Energy, Washington, DC (2021).

[3] UNITED NATIONS DEPARTMENT OF ECONOMIC AND SOCIAL AFFAIRS, 2019 Energy Balances, United Nations, New York (2021).

[4] INTERNATIONAL ENERGY AGENCY, "World Energy Balances (Edition 2021)", IEA World Energy Statistics and Balances (database), https://doi.org/10.1787/45be1845-en

[5] UNITED NATIONS DEPARTMENT OF ECONOMIC AND SOCIAL AFFAIRS, World Population Prospects 2022, United Nations, New York (2022).

[6] OECD NUCLEAR ENERGY AGENCY, INTERNATIONAL ATOMIC ENERGY AGENCY, Uranium 2020: Resources, Production and Demand, OECD Publishing, Paris (2020).

[7] INTERNATIONAL ENERGY AGENCY, Nuclear Power in a Clean Energy System, IEA, Paris (2019), https://www.iea.org/reports/nuclear-power-in-a-clean-energy-system

[8] INTERNATIONAL ENERGY AGENCY, Net Zero by 2050: A Roadmap for the Global Energy Sector, IEA, Paris (2021), https://www.iea.org/reports/world-energy-model/net-zero-emissions-by-2050-scenario-nze

[9] UNITED NATIONS DEPARTMENT OF ECONOMIC AND SOCIAL AFFAIRS, Statistical Yearbook, United Nations, New York, (2021).

[10] INTERNATIONAL ATOMIC ENERGY AGENCY, Nuclear Power Reactors in the World, Reference Data Series No. 2, IAEA, Vienna (2022).

[11] UNITED NATIONS DEPARTMENT OF ECONOMIC AND SOCIAL AFFAIRS, International Recommendations for Energy Statistics (IRES), Series M No. 93, United Nations, New York (2018).